Machine, Meet Human

To my wife Kaylin,
my son Connor and my daughter Rowan:

I love you and thank you for your patience and support
while I was researching and writing this.

Machine, Meet Human
Designing a Useful Interface

nathaniel o'shaughnessey

Meet the Team

Application

"It seems that perfection [in design] is attained not when there is nothing more to add, but when there is nothing more to remove."

"The machine does not isolate us from the great problems of nature but plunges us more deeply into them."

Translations from: *Terre De Hommes,* Antoine de Saint-Exupery

Machine, Meet Human

Why this Book?

The term *User Interface Designer* doesn't mean much to many people. If you're reading this however, it probably does to you or at least it will soon. There are only a few of us out there doing this full time in process control applications. It's a unique job in a specialized application where not very many people end up. Most designers work on it as one of the many tasks they have to work on, not as their primary responsibility. If this is you I'm glad you're reading this book even though it's easier to dive in and just crank out some pictures and get on to your other work. Often there are not experienced interface design resources on-site to help with designing graphics. If there are, seek them out, it can be more involved that it seems on the surface.

Occasionally I am asked to teach people how I design process control interface graphics, what my approach is and where I start. How does one look at equipment, then look at a blank sheet and create a useful graphic? It can be one of those things that seems so simple that it becomes difficult. Too much freedom can be paralyzing, like sitting at a computer with a blank file and trying to make it into a book or article. You have complete freedom but humans work best with restrictions. We do not function well with too much choice. Graphics design can be the same way, that's often why we often seek out templates; not just for uniformity but to limit the choices we have to make. I am writing this to help you in making that seemingly simple leap from nothing to something.

Personally I have been responsible for building several thousand graphics in a dozen plants at a multinational chemical company. Occasionally, other plants ask me to just go over the basics with whoever will be designing them and explain to them how I approach graphics design so they can build graphics for their own plants. This can be more difficult than it sounds at first. I have personally spent more than ten-thousand hours directly drawing process control graphics. With a lot of independent research, reading, internet research, experiments and prototyping in that time. In that time I have acquired a unique education not entirely academic, technical, theoretical or practical; rather a mix of learning avenues. Anyone that has practiced something unique for any length of time realizes how difficult it can be to hand off a summary of everything you know to someone else in a half hour or half day meeting. So I did what anyone without any teaching skills does when asked to teach; I looked for an alternative. Maybe a book for me to recommend that sums up what I have learned on the topic. What I found is that there are a few books out there on to topic, some very good one's even, just none that were really targeted at what I wanted to accomplish. I wanted something that would teach other normal people how to do what it is I do, regardless of their background.

After finding there wasn't a reference to just send people too, I decided I could train many more people by writing a short book instead of trying to train everyone one on one. It ended up a little longer than I had initially anticipated, but it should do the trick. I'm not always immediately sure of all the reasons I make each design decision without stepping back and thinking about it My thoughts and design processes are not always clearly organized in my head, as is the way our minds store information. So instead of wasting people's time, I decided to document my techniques and thought processes and compile them in an organized fashion for people to be able to read on their own time.

It is critical with any task to know some of the background and understand the what you're about to dive into to be able to maximize it's effectiveness. You could be designing for months or years and I would be doing you a disservice if I were to just ask you to watch me draw or to have me watch you draw and critique your choices. I want you to feel confident and empower you to take the lead and make the project your own. Put your name on it and build something you're proud of. It is important to learn to be proud of your work because of its real value; not just about how impressed people are with it.

People that   design control system interface graphics come from many backgrounds. I wanted a book written for an audience beyond just chemical engineers and drafters. Think of it as a crash course in getting up to speed and able to draw interface graphics effectively and independently. My intention in writing this book is to teach some core principles in the most effective interface graphics; regardless of professional experience, platform or drawing package. It's not my intention to go over control room layout design or hardware technologies, nor to teach any process control programming or plant management ideas.

This book should have some useful information for anyone that has to design a computer interface for communicating with humans in any environment. That said, the language and application in this book is primarily targeted for people designing graphical user interfaces for automation systems, particularly for high data to user environments. These could be chemical processing plants, petrochemical refineries, energy production plants, rail system controls or anywhere that efficiency of use is paramount over aesthetic appeal.

There are human factors engineers, cognitive psychologists and other experts in the field that are certainly already capable of designing effective interfaces; however my experience in the chemical manufacturing industry is that these professionals are often not consulted when designing control system interface graphics. Usually the design work is done by a newer engineer or drafter, generally because these people may be familiar with the plant, or at least already in the department and their time can be used for this.

Most often these graphics are designed using vendor templates, a pitfall covered in greater depth later in the book. The other potential problem with this situation is that these graphics are a small portion of the designers work and consequently receives little attention beyond the minimum required to get some pictures up. This usually consists of a few hours, emails and a meeting or two that is spent throwing together some standards based on what they have seen elsewhere or instructed based on vendor palettes.

This is the reality in many environments. The obvious ideal might be to have an additional human factors engineer or cognitive psychologist on staff, but that is generally just not a feasible option financially and is probably overkill. Another problematic situation that can arise is when human factors engineers are brought in as consultants for a brief training. This is a great idea as they should have an excellent understanding of human factors needs and solutions; however, they may have limited experience with this particular application and are generally not available to be a full time designer at the salary a plant would be ready to pay them.

The last option is to bring in a HMI (Human Machine Interface) design specialist as a consultant to help work with existing personnel, who know the process and equipment very well; and enable them to create effective graphics on their own. This is not a bad option either, but they have to do more than just establish some good standards and palettes. They must convey the importance of the work to the specific people that will be doing the actual designing and to the management that appropriates resources for it. They must also teach designers how to approach the graphics, not simply build some good templates to leave, that can be a tall order for a day or week session. The problem with this often is the curse of knowledge; they can take for granted all the things they know, then after they leave, the development stops. No matter how good the start, the momentum must continue throughout the project, then be leveraged to other projects and throughout other plants in the company for it to make a real difference.

In the real world most project managers will have

the engineers or drafters design these graphics. There is certainly a lot of time involved with creating them, as there can be hundreds or thousands of graphics per plant and there is no quick way to create custom graphics. They have to be tailored to the equipment and utilize process knowledgeable people to create effective graphics. This is generally only a minor portion of their workload and they often do not fully realize the extent to which it can affect overall safety and performance. It's not enough to merely be efficient, accurate and thorough when designing graphics, the graphics themselves must perform efficiently, accurately and thoroughly.

Today's rapid expansions and automation implementation requires efficiency in producing graphics, so designers typically look to templates or other plants to model their graphics. This is a logical approach and has mostly been the standard implementation method. The major drawback is that while it is faster, often less than optimum practices get propagated without consideration for new findings and technology. Black backgrounds for instance are mostly reminiscent of when that was the only option. Now with millions of options for backgrounds, black is often still chosen because of it's familiarity. It gets propagated without regard to why it was used in the first place.

Unfortunately, there are still many process safety incidents and a lot of inefficiency that is the result of poor graphics. Graphics that are more intuitive and human centered greatly increase production efficiency and process safety; saving many lives while also increasing profits. This is already known but still many more systems are installed and promoted with less than optimum interface design practices. This obviously isn't intentional but a result of being put together with efficiency of completion in mind rather than efficiency of use. There has been a fair amount of research done, as well as papers written and guidelines published on the topic. Most of these have excellent principles and data to back it up; but still many new installations are implemented with graphics that pay no attention to the findings and do not enhance user operations the way they should.

The core obstacle with interface design is that computers have large amounts of data and process most of it internally; but often it requires a human to interact with it. It is then that attention should be turned to communication of information rather than display of data. Process data is not useful to humans unless it is communicated and understood at the correct time by the correct people. Communication is more than display.

What's the problem? Why are systems being implemented with graphics that are confusing, or more often just plain inefficient? It's not on purpose; it is usually a lack of relevant understanding by the people doing the design, nobody can be a master at everything. They design the way any logical person naturally would. The problem is that we take for granted a large part of the communication we have with other humans. The goal of this book and the graphics it addresses is to help get the useful information to those that need it when they need it.

This book is not a resource intended to aid in choosing a particular hardware or software solution provider. Likely that decision has already been made by the time you are reading this. New solutions companies, new technologies and new platforms will be coming in the future and enable even greater plant awareness and control, but this will bring new challenges in working with humans as well. This book is not intended as a reference exclusively for creating templates or for teaching how to draw or use different drawing or information gathering tools or packages.

This book is intended to give the reader an understanding of what it takes to talk to humans and empower engineering and drafting personnel to be able to make those decisions as they arise and to use existing and new technology to its greatest potential. We talk to humans every day, it is the ultimate curse of knowledge. We take it for granted because it is so intuitive for us. We need to teach that to the machines so they can communicate with us better.

Our computers lack the intuition it takes to communicate with us as efficiently as other humans can. This book is intended to help you design them to work with both sides of our minds. Computers can out pace us in data accuracy and high accuracy mathematical calculations. We humans have a clear superiority is in our intuition and "big picture" thinking however. Teaching machines to work with and utilize our human capabilities is as important as teaching us to work with and utilize the capabilities of the machines.

Up to Speed

Where are we now?

How did we get here?

Alternative

Chapter 1
Where are we now?

If you are new to process control, a new engineer or draftsman or maybe a university student studying process control or interface design, this chapter will give you a brief understanding of what is going on with interface design in industrial applications today. These are the things you can expect when you start looking at vendor templates, previous work examples and existing plant graphics.

Many of the practices discussed in this chapter may not be implemented, encouraged or promoted with every platform, but more often than not this is what you can expect to find out there. There certainly are many exceptions and each platform and each designer has their qualities and characteristics that make them unique and this is a good thing as there are many different applications and teams implementing and running them.

Having varied control platforms gives more choices for different installations and allows us to pick the one that is best for a given environment. This is a good thing and I hope a homogeneous standard platform does not arise as it will limit the more unique applications in the worthy name of convention and uniformity.

You get it; they're not all the same. Let's get into what you can expect to find. These graphics, particularly newer ones, generally look like basic 3D renderings of the equipment in the field. They will often have moving conveyor belts, cutaways of tanks to reveal contents, spinning motors and every color you can think of to use for color coding equipment, backgrounds, title blocks, text blocks, tables etc. Gradients have been around for awhile, but many of the new systems you will find have translucence, smooth gradients, shadowing and many other features that help them look more realistic.

These have come a long way in the past two decades. Thanks in large part to video game design, CAD design packages and 3D animation advancements. That's not to say the individual process control companies didn't come up with any of it, but the millions of man hours that went into pushing the state of the art in those arenas has given the design community many new tools that would have taken centuries for one man or a group to code from scratch. We watch 3D animation movies on TV and they can look incredibly realistic. We design with 3D CAD programs with amazing accuracy and have it rendered so well that sometimes it is hard to tell that it is not actual video footage of the scene.

Hollywood loves this for movies. It's much cheaper to use computers to animate King Kong swinging through the jungle than it is to build a forty foot tall animatronic robot with the lifelike fluid movements to swing through the trees. Hollywood wants to give us entertainment that is often impossible to recreate physically but makes us see it as we see the real world. This expands our imaginations and entertains us.

Those involved with product research and development also love CAD software that enables realistic design rendering. CAD modeling software now allows complex models to be tested and fitted as if in real life then visually rendered to look exactly as a built product would. This allows marketing and aesthetic design professionals to evaluate and give feedback to those in the shop and at the engineering level about changes they would like to see to make a better product, all without ever having any machining, painting or finishing done. This saves huge amounts of time and money. In the product development world timing is everything.

Video games have benefited greatly from computer 3D animations; probably the most like plant control in that it is rendering things in live action and has to display many variables at once giving the player information needed to execute specific decisions. Billions of dollars have been put into developing the technology to render very high resolution live action interactive environments.

There is no doubt that computer rendering has brought amazing technology to our lives. Many of these technologies, ideas and trends are influencing plant automation control system interface design. And that is not all bad; there are some excellent developments that our industry gleans from others.

Video games are often valued on how interactive and sometimes intuitive they are. With games, the goal is entertainment. Often the entertainment is in the game's ability to push the user to the edge of what they can mentally process and force them not to be able to keep up with the objective and thus fail eventually. The user is then ranked on how far their skills can be pushed before they fail.

This is fun for playing a game because it gets our hearts racing, our eyes glaring; we get an actual adrenaline rush. That is fun. Those that master it, have pushed themselves, focused their concentration to its max and practiced the operation to build new mental shortcuts to aid in getting better at playing the game.

For the game manufacturer the goal is to engage and challenge the users, to make it as immersive and attention grabbing as possible. It has to be easy and attractive enough for people to want to sit down and play it without immediately giving up, but at the same time it must be difficult enough to challenge the user and cause them to have to push there senses and concentration to the limit over and over again. This causes an addiction to the game because of the adrenaline rush and the investment of experience into it. Video game creation psychology would be a very interesting field to study; however, since I have no background in video game design or even playing them very much I'll leave that to the millions that have more experience in the field.

There are many parallels between video game design and HMI design. The one thing that is critical to grasp is that the ramifications are real, not virtual. Peoples' lives and billions of dollars are at stake. The temptation when designing user interfaces for these systems is that you will want to make them look nice, look impressive, look life-like. It is a normal temptation to want to strive for that. Because that is impressive we feel more rewarded designing graphics that are realistic. That's a normal inclination and is not always necessarily a bad goal, just not in this application.

Many interface design books, courses and examples are for industries like websites, mobile phones, atm kiosks and other custom interactive screens. These interface screens generally serve two purposes. They first have a function to achieve, but the other factor for those applications is attractiveness and engagement. Certainly in their goal of functional use they use similar human factors principles that we use in process control interface graphics. For these consumer product interface applications there is one more human factor at play that trumps the others. They are selling something and they must create in you a sense of "Wow, that was easy to use, but it was impressive as well, I want to go back there, or I want to buy that."

Back to the state of the industry; process control is a growing industry, which will continue to expand faster than there will be experienced people to design for it. Many principles we cover in this book have been known for some time. Still the growing trend of control graphics mimicking those of other similar computer graphics design fields is the normal scenario.

Each software company is trying to out pace the others by utilizing the new developments from other industries and applying them to control systems interfaces. Leveraging advancements from other industries is what most innovation is really about. Leveraging advancements has led to faster progress and greater efficiencies across many industries.

There are two general approaches to designing interfaces right now. On one end of current designs we have graphics that come from the idea of trying to look as much like the real world equipment as possible.

These are the graphics most modeled after the web and entertainment industries that we just discussed. This is probably the area where the most growth has taken place with new companies releasing new products all the time. This is the bulk of new growth industries. However, there is another older and more broadly applied design practice. You are all probably familiar with reading schematics, maybe even drawing them. This is what early plant graphics looked like. Still you can often see the remnants of that layout. Walk into many established chemical plants or energy plants and look at their control graphics. You're likely to recognize standards similar to that of Piping and Instrument Diagrams or P&IDs. P&ID's are designed the way they are for a reason. While software packages have made them much easier to manage and generate than old mechanically drafted ones, they still retain much of the original aesthetic. For schematic diagrams they are quite useful. Often times since P&IDs are used as the process diagram for designing control system graphics; the graphics end up looking similar to the P&IDs only with dynamic number values and different color schemes.

These are the two major schools of practice for control system graphics design.   Designs based on P&IDs are faster to deploy than realistic 3D renderings as they can essentially be recreated on the HMI with little design effort invested, albeit not without work. It is easier for central engineering and drafting departments to deploy these as they can mostly just look at the P&IDs and create functioning HMI graphics that are accurate and thorough from a process standpoint without understanding the actual process or tasks. Realistic 3D renderings look fantastic and are great for plant tours and showing project managers. Plant operations even warm up to them much faster too, since they are more engaging and have the coolness factor clearly in their favor. Rock star interface designers can be born.

For 3D rendering the extent of design involved usually consists of knowing what the unit actually looks like in the field and recreating that using the latest 3D CAD tools, templates and palettes. Still this takes a fair amount of time and the more time invested and the more advanced the software the more realistic the final graphic can look.

If you are going to implement a control system

interface these are the two major routes currently being practiced. Both require accuracy and both have those that are used to them. Typically larger, older plants have the P&ID style and smaller or newer plants have the realistic style, but you might find either just about anywhere there is a SCADA (Supervisory Control and Data Acquisition) system involved.

Chapter 2
How did we get here?

I've heard it said that when you want to find out why something is the way it is you just have to follow the money. You might think interface graphics design wouldn't have anything directly to do with money, but it does. It's not an elaborate conspiracy to manipulate users into voting an evil villain into office so they can rule the world. It's not that sinister it's just business; but that doesn't mean there isn't any harm done.

Every year process safety incidents occur all over the world in many industrial sectors from chemical and petrochemical processing plants to energy and utility operations and many others. There are thousands of plants located all around the planet in nearly every country on earth. These plants employ millions of people and very few people on earth are out of harms way from some form of a process safety incident. These incidents are the cause for many injuries and fatalities every year. In addition to direct human catastrophes, process safety incidents are the cause of many chemical releases and explosions that do significant damage to the environment as well as costing businesses hundreds of millions of dollars.

A true tragedy was the incident I'm sure anyone reading this book is familiar with at a Union Carbide co-owned plant in Bhopal, India in 1984. This runaway reaction and resulting gas release killed thousands and injured hundreds of thousands. Up to half a million people's health was directly affected from this incident. What's worse, this incident has been directly attributed to human error. There were many factors involved from negligence, poor design, horrendous maintenance and procedural practices, under-trained operators and even local officials neglect. Financially, it eventually sunk even the parent company Union Carbide. This doesn't even take into account the affects to the environment of the release of forty tons of Methyl Isocyanate gas. Countless animals and plants were directly killed in the release along with the human casualties. This was a horrible accident that most certainly could have been prevented. If they had today's technology with high efficiency process control interface graphics along with better training, it could have been avoided. That doesn't mean something like that couldn't happen again though. Poor interface graphics design and poor plant operations practices still exist today in many places.

More recently an explosion in Texas City, Texas in the U.S.A. in 2005 at a BP refinery killed fifteen people, injured one-hundred and seventy people and cost BP well over $1.6 Billion in total. These are extreme examples but real examples; very real examples. These are just two of the biggest process safety incidents in addition to the recent oil spill in the Gulf of Mexico in 2010; but many more incidents where people's lives are lost and people's health is lost occur every year. Even more often than that, there are many incidents that cause damage to the environment or cost the company money in lost material, clean up costs and loss of production.

Often, these incidents are due in part to interface graphics that failed to inform the operators of critical information. Many times information was available or even displayed but actions weren't taken in time to prevent an incident. This is usually indicated as a human error. Sometimes that is the case; but if the operator fails to understand something displayed on the graphics, this is more accurately a communication error. It may seem a trivial difference. After all it is the human's job to monitor the plant and take appropriate action to avoid an incident and to react to an incident. Ultimately however, the responsibility of the graphics doesn't end until the user clearly understands the information needed. It is simply not enough that is was displayed. We'll get into more of the mechanics and dynamics of an effective interface later. Understanding that the cost of process safety incidents is enormous is critical to understanding the need.

Back to the money trail. It's clear that huge amounts of money would be saved by having these interfaces be as efficient and as effective as possible, right? Yes, having them be as effective as they could possibly be at communicating what they need would be the obvious money savings and be the most efficient for all involved. It seems like an obvious answer, the money should be incentive for optimization.

But that is skipping a few steps. Let's follow the

28

actual transactions and decisions. First, a company that is going to convert control systems or build a new plant will decide what type of control system they believe will work best for them. Maybe you are in this situation right now or have just finished going through it and are starting to design the graphics for the interfaces.

Sometimes graphics packages are included with the control system platform and some times they are from a separate company as an add-on or for remote monitoring. In either case the vendors that are selling these platforms have already put in huge amounts of effort, time, work and money to develop them. They do not get any money for their work until someone buys them. That is a lot of pressure to make a sale. The existence of the company often depends on selling these highly sophisticated, complex and expensive control systems.

These control system vendors have to put their best foot forward. Reputation and system capabilities sell platforms but so does the sales pitch. There are many factors like IO count, cycle rates, architecture design and programming language as well as compatibility with other components, technical support and other factors. There is also the HMI. This small portion of the product that can make or break a sale and possibly in turn, the company.

The HMI is the face of the control system. It's what people see when they look at a system in a control room. Many of the other factors above go into the logical portion of the mind of the person or group responsible for ultimately purchasing the system.

But the other half of the brain also has to be convinced for a sale; the side that needs to be convinced that it's a good match aside from all the specifications. Does it "feel" right? One of the reasons we are in the situation we are with control system interfaces is because often the people making this decision are analytical people who don't think they are affected by such emotional pulls; but we all are. We all must be sold on both fronts to make a committed decision.

Often our one side will sell the other in the absence of a convincing argument from the seller that appeals to it. Sometimes the logical side will tell the emotional side that it is a good thing and you should feel good about yourself for being so responsible as to make that responsible decision; nothing wrong with that.

Sometimes it's the emotional side that is sold on something and will find arguments that can convince the logical side. There are plenty of examples of this in every day life; how we convince ourselves a decision is right for us. It is completely normal and it would be difficult to function otherwise, but we must be aware of the process.

For a comparison let's look at another more familiar industry; the automobile industry. A buyer must make a committed decision and act on it for a transaction to happen. For a sale to occur there are two sides that have to sell. The first is what the car can do logistically, its specifications, but there is also the other side. What does it look like? do I like how it looks? Or more accurately sometimes, how would other people look at me when I'm driving it. We'd like to think we're logical and we would only buy a car based on utility, efficiency and other logical inputs; but do we?

Why does every car company spend a large portion of their R&D budgets on making a car appeal visually or to our sense of identity? Even hybrids and utility trucks are advertised this way. We might think we are making that decision logically but every marketer knows the clear understanding that there are two portions of our mind that both must be appealed to. Marketing is almost as active with these vehicles that are supposed to be logical purchases as they are to sports and luxury cars that are more open about their appeal.

Utility trucks are utilitarian; but why bother with all the chrome, nice paint and decorative accents? It's because to drop that kind of money, we have to like something, not just decide whether it's the most efficient or not. That's human nature; we all have it, myself included. We just don't always recognize it.

Back to control system sales. There are few indicators for the emotional side of our brain in a control system sales presentation. The HMI, the face of the machine, is one of the few direct appeal avenues the vendor has to sell you the system and they have to do their best. This isn't their fault alone though. If people were more conscientious and logical buyers, the vendors wouldn't have to put any effort into making the HMI as visually appealing as possible. We humans are all that way to a degree and we need to be aware of it.

The HMI screen shots that are presented on the company's website, displayed in a presentation or installed for a demonstration have to be as engaging as possible. People have to see them and have that feeling of "wow" in the back of their mind. They are going to spend millions or hundreds of millions on this control system. You better believe they need to sell to every aspect of the buyers mind. If they see the HMI screens and think to themselves "this doesn't look high tech, sophisticated or advanced, is it really as good as they say?" They are about to spend millions of dollars on a system and they need reassurance they're getting the best option for them. They need to feel it's the right fit for them; a successful, complex, sophisticated and state of the art plant. The screens need to reflect that. If they don't, the buyers' minds will look more at the negatives and ignore the positives. Likewise, if a buyer is impressed with the face of the machine, they are more likely to overlook slight technical shortcomings and look more at the positives.

Just like an automobile or mobile phone sales pitch that is partly about sheer visual appeal to the buyer; it also needs to appeal to the sense of identity of the buyer and how others will view them. Does this car or phone show others that I'm practical, important, smart, capable, efficient and responsible? The same goes for control systems.

This appeal plays an integral role, albeit probably not the main role, but with everything else comparable, the vendors must use this smaller portion of the sale to tip the decision in their favor. On the logical side they may be neck and neck going into the mind of the purchasing plant. In this situation the interface example designs can make the difference in which company gets paid for their work and which company is in a tougher position at the next sale and the next after that.

It's not malicious, it's sales. They are not trying to set plants up for failure; they are trying to stay alive as a company themselves. After all, they are generally made up of software programmers, hardware engineers sales, management and support. They have others on staff and have a diversified team, but their goal is to make and sell the best control system there is. That is the purpose of their existence as a company. This dictates every decision everyone in the company makes.

Once again I'll return to the example of the automobile because it is a similar market structure. It's a piece of equipment with real value, not just a luxury.

The automobile and all its marketing avenues are part of our life and provide an easy reference we will all relate to and understand. When an automobile manufacturer wants to sell you their car there are many types of commercials, but they usually don't depict real world scenarios. Sometimes they will have a brand new fully loaded version of a vehicle driving somewhere nobody ever actually would in their daily use. Here's a staple: A car is driving around on the salt flats sliding sideways at eighty miles per hour in slow motion. Then there are the SUVs that are driving around on undisturbed snowy knoll going who knows where. Don't forget the pickup trucks that show a giant payload being dropped into the back of the truck and then driving over big boulders through the woods. Bye-bye fancy paint and I hope you didn't want that oil pan gasket to hold after it's smashed on the rock. Sure, maybe these vehicles can do these tasks they show, but that doesn't mean they should.

This is all about the sales pitch. We all know it. It doesn't show us what it is intended or engineered to be used for. Most of the time it depicts a driving condition that would void the factory warranty should something break while doing it anyway. They certainly wouldn't accept the blame for the damage that might occur when the car hits a hidden ledge or drop off under the snow and completely ruins the vehicle.

We all know those are exaggerations and we accept it as part of the marketing. We realize they are simply appealing to our emotional brains' sense of adventure and freedom and we're OK with that. It doesn't mean we go buy one and mimic what we saw in the commercial.

We don't blame the car commercials when someone does actually drive that way and somebody gets hurt or damages property. We are expected to use our own judgment on the proper use of the product we purchase. Thankfully there is an abundance of material and training on the proper use of automobiles and there is a universal understanding of the proper way to do it. It is understood that it is often not the way it is in the commercial or the magazine ad. When someone actually drives that way and causes a driving safety incident it is not the car companies fault, or at least not entirely. It's the way the marketing system works and if people aren't aware of that bad things happen.

Likewise, control system vendors are not entirely

to blame for the graphics that are used in the plants being over complicated or too flashy for their own good. They are merely selling the product, how it is used is still the responsibility of the users.

This explains where the high visually stimulating graphics come from. They are created by programmers that are hired by the software company because of their expertise in complex programming and state of the art knowledge of the capabilities of the software and hardware available. Naturally, they want to show their managers what they can do, so they churn out some great looking and complex functioning graphics to be added to the control system. This ends up getting pitched with the system. With all other variables comparable the most impressive looking system makes the sale and that company survives. It's almost reverse-Darwinian how we got here. Thankfully this doesn't always happen and isn't the way every vendor works, but it is fairly typical.

Some control system vendors actually even do the design of the control system graphics for the client. This is not the most common, but it's often added as a sales feature to help set them apart. This seems like a streamlined approach for the plants because then they don't have to worry about any of that. Leaving it to the professionals is often a good idea. The problem persists with this scenario however also. They need to deliver something impressive to the plant for the plant to be happy with it and choose to continue their services. Helping land them future contracts and keep their business alive and getting paid; simple business again.

Scenario two is more common. Scenario two is when the vendors deliver all the hardware and help get it set up. They will sometimes provide templates of different styles for the plant to choose from. Templates are almost always given when purchasing software like this and having the templates often leads people to believe they have to choose one of those options. The vendor was really trying to show some of the diverse capabilities, not make a statement of the way it has to look. They have to provide a range of options, but they all have to be impressive to be taken seriously by potential users.

Just as Vincent Van Gogh painted many different paintings, they still didn't look like a *Dilbert* comic. Sometimes *Dilbert* can be more straightforward and easier to understand albeit much less impressive or moving. Getting multiple opinions all from the same point of view may give you different tone, language or aspects; but it's still coming from the same place and motivation. Employees of the platform vendors go to work for the same reason we do. Their motivation is making money and hopefully building pride in their work. Pride and money are both gained by the plants being impressed with the provider's product. So, that's what the plant gets, what the plant wants.

An engineer that is working on a control system retrofit, upgrade or new plant installation may often choose to use the same system that is already implemented in other plants within the same company or sector. Cutting down the learning and adaption curve; a great idea. Often however, the same standards and practices are used that the previous plant used for the sake of uniformity. The idea of "why reinvent the wheel?" comes to mind; regardless of the fact that often the previous wheel was inefficient or was designed for the previous vehicle not the new vehicle being used. There is certainly nothing wrong with uniformity and leveraging previous work and expertise. In our quest for efficiency, predictability and uniformity we must be careful to still be considering the what the best way to implement that same system is. We need to continue to view design as a dynamic process that constantly needs to be re-evaluated and not mindlessly designing new graphics that aren't really ideal for their new environment. Just remember to keep thinking.

The job of actually doing the drawing for all these graphics, sometimes numbering in the thousands for a larger plant, is tedious and often viewed as semi-mindless requiring very little training or experience in the field. It is sometimes viewed as a technical task and is given to someone to basically recreate the P&IDs on these graphics. Other times it is viewed as a creative endeavor and given to the most colorful or creative engineer [yes, I realize the irony] or a graphic designer is pulled in or contracted for this work. Usually it falls in the lap of the youngest engineer, a draftsman or intern. After all, these types of projects requiring complete control system graphics design are usually huge projects like new plant construction or control system retrofit.

The complexity of what goes into a project of this scale is mind boggling and the more senior engineers have tons of work and usually are already working overtime just to get the engineering ready in time and implemented in synchronization with the whole project. It truly is an insanely complex task.

With everything going on in this time, whoever gets the task of cranking out all these pictures is often one of the lesser experienced in the group or someone from outside the group altogether. That's not a bad thing, that's how I got my start in the field, but it does stack the cards against you in terms of getting it right the first time. I know I didn't. After all, the "right" way is not the most impressive looking or the most interesting to design. The odds are not in favor of these being as effective as they could be.

In the rest of the book we'll try to look at what a better way actually is. The trouble is that the most efficient way is also the most boring way. Even when someone does do them the best or "boring" way, those never seem to catch on. The next person that comes along and looks at them doesn't realize all the work and the principles behind why they are the way they are. These new designers are fresh out of school, or coming in from web design or some other interface or drafting background. They're hard working and eager to prove themselves and they don't realize that this seemingly boring style actually does make for a safer, more efficient plant even though it doesn't look as impressive.

Accepting and just reworking those boring graphics the last guy put together doesn't seem rewarding. They either revert back to vendor templates or search online for HMI examples. Sometimes the goal is just plain getting the job done as quickly as possible to impress with speed of completion, but at the expense of thought and research. Sometimes because a project is on a time line and the designer has many other jobs to finish and just getting out something functional is usually expected or delivered regardless.

At any rate, even when good practices do make their way into a plant, it is likely some of those practices will be abandoned in favor of more industry typical, highly visually stimulating graphics more like the one's we have been discussing. This is the state of most of the industry today and the reason for this book. How can we change it and equally importantly what changes are needed?

Chapter 3

Alternative

We have discussed the two most common design styles: P&ID based and realistic. There is a third idea however that has been around for decades, but doesn't seem to gain real traction despite its proven efficiency.

This alternative option is not based on technology, either new or old, but on people. It starts with understanding people and how we see things and tailors the graphics to work around our minds, instead of the other way around. Most of the people promoting this idea are academics in cognitive psychology, but still it doesn't gain traction. Primarily, I presume, because of the reverse-Darwinian market dynamics covered in the previous chapter.

This third group is an alternative idea in comparison to standard practices. It is considered by some to be a waste of time and to others it's considered

43

to just be academics being overly-analytical with their psychology mumbo jumbo. It is primarily human factors engineers and cognitive psychologists advocating its use and that may have something do with the stigma and why it isn't permeating the majority of plant implementations.

This human centered design is not really that complicated and it is not really that far out there. It does require a shift in thinking for some though. And that is the real enemy. Those in the field of human centered design may find some of the ideas and principle easy and elementary; but to the rest of us it seems boring, dull and maybe just a little like lazy designing. It is not about a higher level of understanding or knowledge, or a highly skilled programming language or software application. It is a change in thinking and sometimes that can be harder than learning entirely new things.

The basic idea of human centered graphics design is that data is meaningless without context and context is just a picture without information making it useful.

Situational Awareness is the term often given to the act of driving the plant rather than running a plant

44

based on reacting to the computer or by manually commanding the computer. It is more like a commercial airline pilot who works with the plane and all its complex systems and automatic functions. The pilot neither commands all of the functions of the plane nor is spending all their time reacting to the alarms and flashing lights in the cockpit. They have to work with the plane to use its sensors and automatic functions as an extension of their mind and body.

Think of a Formula One racing driver. They do not try to command the entire process. They do not stop after each move and think about what they should do next and then execute that decision then analyze the new surroundings and watch for changes then repeat the process. Also, they do not just start going around the track and react to everything. The racer doesn't wait until they are rubbing the boards or bumping another car to brake or turn. They do not just put their foot on the gas and wait until the engine overheats, red-lines or plateaus to shift gears.

They are not reacting to the environment nor are they controlling the environment. They are working with the car to drive incredibly fast and make split second decisions to avoid negative consequences that may cause an accident or lost time. They do this in an ever-changing environment with too many variables to consider. This would be incredibly difficult to program an automated system to master alone. At the same time it would be incredibly difficult for a driver to do without using any automatic systems. If they had to manually control the fuel/oxygen mixture along with many other dynamic functions while being aware of everything else around them it would be impossible. When the driver has optimal situational awareness is when they can drive at an incredibly fast, but efficient and usually safe pace.

This is the idea at the core of human centered process control interface graphics design. Operators shouldn't be running a plant by reacting to alarms and flashing lights. Nor should they be making a decision, issuing a command, waiting for the changes, then deciding what to do next. It is most efficient when it churns along like a finely tuned car and a relaxed but focused driver. The driver doesn't have beautiful 3D renderings of the engine, car and track in front of them as they are going around the track. Also, the driver doesn't have a table with all the numbers that the computer is calculating on the dashboard. Both of these would be a distraction and a waste of visual space and mental processing capacity.

Ideally only the minimum context is needed to provide the least amount of mental processing to gain the appropriate information and mental understanding of the current process. It's the idea behind analog gauges on passenger vehicles still being the norm despite the fact that digital gauges would be more accurate, easier to install and look kind of fancy too. The analog needle gauges in most cars now are usually reinterpreted from digital signals with stepper motors anyway. Why do they go to that extra effort? It isn't just nostalgia from an era before computers were used in automobiles. It's because the analog needle gauges puts the data in context giving useful information to the driver out of the corner or your eye without even having to focus on it.

It's the idea of situational awareness. Things that need your attention like seat-belt lights, door ajar, check engine and so on are shown using a red or orange symbol or light. Even the trend towards white or more neutral gauge backgrounds is based on the principle of optimizing mental processing. That's not to say automobile dashboards aren't designed without emotional appeal, but they are one area that human factors engineering has gained a pretty good foothold in the design.

How do human centered graphics differ from other conventional schematic and high resolution graphics? We'll start by first analyzing some of the ways we humans see things, understand things and make decisions. Also we need to understand why and how computers see and make decisions.

We spend a lot of time and money teaching people how to do their jobs and machines how to do theirs. We also spend a lot of time and money teaching humans to communicate to computers in their fashion. The missing link is teaching computers to work with the way humans communicate. After we explore this we can see how to create human centered interface graphics that are designed to help the computer work with the human operators.

Meet the Team

Understanding Computers

Understanding Humans

Communicating with Computers

Communicating with Humans

**Teaching Computers to
Talk to Humans**

Chapter 4
Understanding Computers

Computers do not know when something is abnormal unless they have been programmed to calculate it. They don't see obvious problems unless they are programmed to recognize a set a variable states that indicate a problem. Computers have come a long way and continue to develop in capacity, speed and efficiency at an incredible rate and in all likelihood will continue to increase in processing power and memory for the foreseeable future. Likewise, software will continue to build on itself and as well and it's complexity will continue to rise accordingly.

Computers are only able to think in a logical way and this is incredibly useful for processing large amounts of data and crunching many calculations in real time. Computer logic is based on its human programmers and is in a way a reflection of our own logical brain functions.

Accuracy of memory is something computers

51

excel at as well. Humans have an incredible memory and useful organization of memory, but our memory is not always accurate. Sometimes we forget things altogether and often more damaging, we occasionally remember things inaccurately, developing spontaneous memory sometimes based on real life, but often altered by interpretation, dreams, suggestions, reconstruction, meditation and other things. A computer's memory is accurate. It may get damaged, but it is rarely corrupt in that it will remember incorrect information. The statistical odds of the logistics of that are very slim; especially when compared to our human memories.

In the processing portion of computers the operation is completely digital. Analog processors are in development but it will likely be a long time before they are used commonly. Analog IO are interpreted for the digital processors by way of rounding, sometimes to great accuracy, but the logic is always digital. This means at its core a bit value is either a true or false, there are no gray areas or room for error when processing and making decisions. Computers do have apparent glitches and don't seem to do what they are supposed to. This is just an illusion though, sometimes

52

the calculations are just too complex for us to follow and it comes out with a behavior we didn't predict. However complex they always do what they are programmed to, even if the programmer doesn't fully understand the possible outcomes.

It may be a hardware issue causing them not to function properly or it may be programming that didn't account for every possible scenario.

In all fairness to hardware manufacturers, it is impossible to build hardware that will never fail. Entropy is present and with thousands of components that need to work together there is ample room for error due to malfunctioning hardware. Before we blame hardware manufacturers think about everything that is involved at the most basic level in creating a computer system that can process thousands of hardware signal I/O. Not to mention the computation and display required. If you're like most of us, we don't even know what goes into everything at the basic level, but we expect it to perform flawlessly. Personally I am very impressed with how well the hardware is manufactured considering everything it has to do. It is possible that some component may fail somewhere and the system will keep working, but will not process it the way it was intended because it is receiving wrong inputs or the outputs are malfunctioning.

On the same note programmers should be given some forgiveness as well. Whether source code, machine code or end use application programming. The more variables there are in a system the more complicated it becomes and the harder it is to contemplate every possible scenario. So when the computer tries to calculate a scenario it has not been programmed to calculate or rather consideration wasn't taken to account for every possible variable it can have seemingly erratic behavior. It still is only doing exactly what it was programmed to do. The apparently erratic behavior is just that proper consideration wasn't taken by one of the levels of programming that went into the final product for every scenario of variables including the possibility of hardware malfunction of one of the thousands or millions of components.

Let's do some quick math. With 8 digital variables there are 256 possible combinations of outcomes that all must be accounted for to have a completely predictable outcome from a calculation. Now figure many plants have 1,000 straight digital hardware inputs. Now there are 2^1,000 possible combinations assuming all the hardware, processors and other components are working as intended. Assuming you know your exponential math you already know nobody could ever account for every possible combination individually. Now throw in thousands of memory variables that can be factored in and thousands of analog variables as well then you get the picture that the outcomes are nearly infinite.

The way we organize the potential scenarios helps us manage it and we generally can achieve a predictable outcome most of the time. There are many programming languages and shortcuts that attempt to manage that seemingly impossible task by grouping variables to exponentially reduce the possibilities and many other common programming techniques. It is still very possible that not every potential outcome was anticipated and programmed for and then we can get seemingly mysterious outcomes.

In the end though as long as the hardware always behaves in its intended way computers always do exactly what they are told.

Chapter 5
Understanding Humans

Computers are similar to our logical minds. The part of our minds traditional economics assumes makes all our decisions and the part we use for much of our mental processing functions. We look at all the variables and make a calculated decision on what we should do. We do it constantly during our waking hours and often while we sleep as well. Our logical brain is constantly analyzing all its inputs both environmental and from memory. It is making calculations and outputs to our muscles and nervous system as well as to our memory and saving values for our next calculations. In that manner our brains are incredibly computer-like and many would argue we still rival computers in that regard. We just take for granted all the subtleties and endless variables we encounter.

Ever think about walking? Neither do I. At least not very often and not to its true depth. Some part of our brain is thinking about it and calculating it continuously. Up to forty times per second or more our brain will analyze things from short and long term memory as well as current environmental variables. Variables from memory include things like where are we going, how fast are we going there, what type of stride will we use. Not just for speed but in relation to others observations or how we "feel". Do we stroll, walk or strut? Are we carrying something fragile that we need to be extra careful with? Are we walking with someone we have to monitor and adjust pace with? Are we holding our child's hand and have to factor in things like slouching to the side while we walk or do we have to provide additional support should it be needed? That is just scratching the surface of the variables that are pulled from memory that must be taken into consideration during this calculation.

There are also other factors that impact that calculation as well. There are inputs I refer to as "network" inputs, which are signals from peripheral brain functions going on simultaneously in another part of the brain; think of a computer network or sub processors. It receives variable inputs from our balance system which has all its own calculations to manage. Think of that as a distributed control system, it is doing all its own calculations all the time and outputting to other parts of the brain. The walking portion is receiving the variables as inputs that the balance portion uses as outputs. Things like: are you leaning forward or backwards, are you spinning one way or the other and so on. The portion of your brain controlling the walking has to take all these network inputs into consideration while calculating the next set of outputs.

There are physical and environmental inputs that are considered as well. These are also nearly impossible to list completely. I'll start though; just to get your brain started thinking about it. We have to consider what position everything is currently in. Like what exact position is our leg in right now, what angle is our foot at, where is our center of gravity, what position are our arms and head in, as those also affect our stride. Even things like our breathing and gum chewing are inputs that could affect our calculation of what set of output signals to choose. Then there are signals from our eyes. Do we see something that requires immediate reaction? Do we see that the next step will be lower or higher or off

59

camber?   Do we see something that will change the course we need to take?   Maybe a puddle or stairs. Then there is our hearing as well.   Do we hear something that would alter our path in some way? Also our sense of touch.   Did we stub our toe on something and now we need to adjust?   Did we make the wrong decision previously and now we feel more or less pressure on our foot than anticipated and now this next decision and action needs to account for that?   Those are just a few of the environmental and physical inputs that are calculated in our next decision.

After all this information is gathered by our sensory and network inputs, our brain will make a calculation as to what signals to send out for the next cycle to achieve our task of walking.   This is more of a dynamic function split among many different processors cycling at different rates and different levels of complexity.

The calculations and outputs are as complicated as the inputs were.   How much force and speed goes to which muscles?   Is there anything that needs to be put into memory or outputted to other network peripheral functions?   Do we need to turn our head to see something we heard from our last set of inputs? What position is our head in now? How fast and with what force does it have to move to get where it needs to go when it needs to go there? You can see how there are almost innumerable outputs that must be sent as well to keep us on our intended course.

All these calculations happen many times a second in our brain for us to achieve fluid movement. Fortunately for computers we modeled many of their calculations the same way our brains do, using shortcuts like variable packing. Our brains form their own shortcuts through practice and observation, some of these shortcuts we refer to as reactions and auto-piloting. Another shortcut is the idea of change. We can do things like have a set of anticipated values for each of those values that are analyzed. Often just using whatever the previous value was or what the anticipated value based on the previous cycle's calculations and anticipated change.

Think of the basic equation: current value minus anticipated value. If they are the same it equals zero. Sometimes there are hundreds or thousands of variables that are all as anticipated and then they all equal zero.

61

They can be added or multiplied and the new value also equals zero if everything is as anticipated. This allows little changes to be made to the anticipated set of next actions even though there are many variables considered. This can allow one portion of the brain to calculate the anticipated return values and do its operation and output it as a single value to the part that is processing the action of walking. When everything doesn't go as anticipated adjustments are made for the next processing cycle to compensate for the difference from expectation to bring the intended outcome back into line.

There are many other ways to look at it and many other shortcuts but we do not really need to go into all of them. Because, quite frankly, I certainly do not know them all and I am certain the rest of scientific community does not have them all mapped. My previous observation may or may not even be completely accurate as to the logistics of the operations of the shortcuts we use. The point is that we need to notice that we use many complex filters and shortcuts to manage the infinite amounts of data we have at our disposal and we clearly cannot think about it all. We practice things and build complex reactions and anticipated reactions to our actions. The scenario we examined at a high level was a very basic function. Consider a gymnast or basketball player, or even a musician or actor. Even many of the functions we perform day to day that are not as physically complicated but still require extensive observation and computation, like reading someone's body language as we talk to them while we're walking.

Computers often work in a similar fashion except they have to be programmed to do so by the conscious mind of programmers. Most computers systems still have to process it all through a central processing unit. Fortunately DCS *(Distributed Control Systems)* are architectures a little more like our brains, with multiple distributed computers, PLC *(Programmable Logic Controllers)* and micro controllers that handle much of it at different integration levels. Our human brain still gives even the fastest, most complex computers a good race in terms of overall data "processing". We just use so many shortcuts and reactions to all the variables we take in that we often tend to dumb down the true extent of what our brains are processing.

This is all dealing with two major portions of the brain, the conscious mind and the reactions trained into our nervous system. Computer systems still do not come close to mimicking the full capacity, adaptability and flexibility of these cognitive systems, but they are built with the same general principles that guide the logical portion of our mind.

The state of machine systems hardware and software is advanced enough already to mimic enough of the logic portions of the mind and variable monitoring, calculating and outputting to control the main logical steps that a current automation system employs and far more accurately than humans can. Due to our complex minds we get unintended outcomes much more often than computers.

Where computers really have humans beat is in variable value storage. We can build computers and write software that can monitor data from thousands of inputs and store and recall them with a very high accuracy. If the hardware and software are functioning properly it is always completely accurate. Average humans on the other hand could probably monitor about three to five numbers that are changing and even then the refresh rate would have to be relatively much slower than a computer and we cannot have an accurate rolling data log of the events either. This is a limitation of the human mind. It is incredibly efficient and complex in many ways but high accuracy data logging and breadth of conscious observation is very limited.

There are many tasks in a plant that we are not easily able to automate and the plants need operators and specialists to perform all the functions that the machine is not capable of. Humans are flexible and our minds and bodies are highly integrated; giving us great utility. We are capable of doing many tasks and handling abnormal situations quickly and efficiently. We are able to analyze situations in a way that would stump a computer system. This again comes back to the idea that every possible scenario has to be programmed for but the number of possible scenarios is greater than any group of programmers could ever anticipate. Let alone construct logic and build hardware and plant equipment to be able to handle even if the scenario was anticipated.

Humans fortunately are equipped with something computers are not. We have another brain function, commonly called "right brain" functions.

There is still some debate about the accuracy or rather the extent of the applications of the term "left brain" and "right brain". It more commonly refers to our "logical" brain in contrast to our "intuitive" or creative brain functions. More logical functions such as mathematics, speech structure and logical processing takes place on the left cerebral hemisphere while more intuitive functions such as reading faces, language tone and creative aspirations etc are processed on the right cerebral hemisphere. Thus the term "left brain" refers to functions that use our logical brain functions. This side is more computer-like than the "right brain" functions which are more artistic, creative and intuitive.

Typically functions of a *left* brain can be laid out with logic. Functions such as math and language mechanics. Conversely right brained functions have a harder time laying out clearly what is going on in a step by step chain of logic. More free association and logical leaps take place in those functions using incredible amounts of data in a way that often the cognitive mind of the person doesn't even realize. You know how when you see someone from across the room you might get a feeling about that person and form an impression of them based on nothing logical or easy to explain. This is intuition or a "gut" feeling.

Behind the scenes it usually has to do with facial mannerisms, body language and vocal queues along with experience from previous interactions with these variables. This is hard to even call a calculation because of the number of variables involved and the fact that we would have a hard time laying them all out and following the logic. In other words we would have a hard time teaching this to a computer. These are brain functions which are far to complicated to logically think about but somehow most people seem to be able to interpret them.

The information and variables are so vast yet people's interpretations can sometimes be universal. That's why there are entire fields of study centered on this concept. However you look at it people have the ability to look at situations and sometimes derive much more information about them than we can program into a computer or easily explain. Many of these abilities are common among the majority of people in our society. This is something the computer cannot really understand but can leverage if it is taught how to communicate with

67

humans effectively.   These are sometimes not signals that can explained with equations.

This right brain is often associated with artistic and creative expressions.   At first it seems it would be pointless to try to teach a computer to have an intuitive mind.   That part is true; we don't want to try to teach a computer to have a right brain, just to be able to communicate to ours.   What does intuition have to do with running an efficient, safe and productive facility? Well, maybe you don't need to start a band in the control room or contemplate just the right mix of colors to create the aesthetic you want for the label on the pipe.   That doesn't mean this additional mental capacity cannot be used by the computer much more than it commonly is. Sure; it might make us more likely to daydream of being somewhere else and that is a safety and productivity issue but it does many other functions as well that can be useful to the process.   Human operators are far more than adaptable robots to carry out tasks there isn't existing equipment for.

Maybe the value of this other half isn't apparent just yet but we do need to recognize that it is there as it works hand in hand with the logical part of our brain.   It is present in everybody regardless of how noticeable it is; I assure you it's there. Humans are not always rational and logical; we need to understand that. We may have logic that can function similar to a computer, but we also have irrationality that works sometimes in opposition to our rational mind.

We have already established that humans are most certainly necessary to the process due at a minimum to a human's extreme adaptability to many specialized tasks. We're the ultimate in multipurpose tools. We have many other qualities and capacities beyond being a biological Swiss Army knife. We are logical and can make very organized rational decisions; but we are also intuitive and can make decisions sometimes that are hard to pinpoint the exact logic even thought he purpose might be clear. We can make leaps of understanding that may not always have a clear path of reason even though they sometimes are so universal that we consider them common sense.

All automated facilities have some level of balance of responsibilities between man and machines. The optimum balance is different from one place to another but the greatest efficiency is in striking the right balance for that scenario and in optimizing communication between the two.

People can learn machines. Maybe not completely; nobody understands every last thing about a complex machine like a SCADA DCS. Just like getting to know another person, however, the longer someone works with a machine the more they get to know how it functions. Like riding a bike or driving a car; it takes a little practice and with some education and experience we can learn to work with it to achieve the intended goal.

We always have to talk to the computer in the language it understands though. It understands logic. We have to talk to it using our logical "left" brain primarily. Input value here, push a button there and open a valve there. Unfortunately, it talks to us only in this language as well. It only generally gives us logic based values and options. Which works alright. After all we do have fairly logical brains and we are able to communicate relatively effectively in this way.

Imagine now the possibility of it being able to talk to us using both portions of our brains; our logical side and our intuitive side. The intuitive side can do things our logical side cannot or at least not as efficiently. It can glance at a picture and immediately know if there is danger or something needing attention right away. Instead of our left brain's technique of looking or scanning through values and comparing them to what they should be to see if something is abnormal. Sometimes also relying solely on our procedural reactions to the computers outputs.

An example of this is running a plant by reaction, as many plants do. The operators do their routine tasks and an alarm from the automation system lets them know when they need to do something. As inefficient as this should seem, it's the way many facilities operate. No driver, just operators reacting to the automated system. This can happen either by intentional design or an operating culture can migrate to this because the computer is communicating to them in such a way that they are unable to use their intuition or even their logic to "drive" the plant. They are forced to just react when they are told. Thus needing a driver seems a waste if they are not really driving anything just reacting.

In a multi-station SCADA environment there can be several drivers that are all driving at the same time. More typical is each driving a different part but working together. Think of the tiller ladder fire engines that have a front and rear driver that work together.

It is important to know that just like in many other environments communication is as important as capacity or skill level. You must be able to communicate in both directions effectively with the people or machines you have to work with to really operate as safely and efficiently as possible.

Chapter 6
Communicating with Computers

Anybody speak binary? Anybody have network ports, serial ports or USB ports? OK, so we all realize computers don't speak our languages directly and we don't speak their language directly yet we still communicate. We use a basic mouse and keyboard primarily to communicate to the computer. Occasionally other means but those are the two primary means and we will assume that's our mechanism for talking to the computer. A lot of research and development has gone into the modern keyboard and mouse even though they may seem ordinary.

Many people are still not convinced in the adequacy of these primitive input devices and are pushing many new technologies. For instance, gesture recognition is making some good headway and I know there is a lot of excitement about employing it in things like HMI's. This idea is thanks in part to it being put into movies as the future of tech. Directors use them because they look cool. I have to agree they look very "future tech." Being a geek at heart I really would like them to be a viable option because they look cool. The truth is that for use by anyone spending much of their time working at a workstation like a SCADA HMI, it just isn't ergonomically viable nor is the most efficient. That is not to say that it has no place in industrial controls. I think there is some possibilities for this technology but not as the primary supervisory control of a plant. This is newer technology but due to things like Gorilla Arm Syndrome it is not a viable option.

Gorilla Arm Syndrome is a condition where a human holding or moving their arms in an elevated position for prolonged periods of time suffers several unwanted side effects. The biggest side effect is clearly fatigue, which is uncomfortable; but also quickly results in lack of fine motor control. For a gestural interface that will cause issues of precision control. This would become frustrating and greatly reduce plant safety and efficiency. It would be impossible for anyone to use gesture controls for hours on end the way operators need to use the interface systems of the SCADA. Due

74

to fatigue issues and a general lack of precision; gestural controls are just plain not suited for this application.

When it comes to talking to the computer it turns out that for now the old reliable keyboard and mouse is the best we seem to have right now.   Direct mind interfaces are still decades or more away for this type of application and I would guess their precision will still not be accurate enough for implementation in a process control system.   A mouse and keyboard are still the primary controls and my prediction is that they will remain the most efficient means of communicating to the SCADA for quite a while.

Personally I think it is likely that their value and efficiency will actually increase in the coming decades. Previous generations didn't grow up with a mouse in one hand and their other laying on a keyboard. I didn't even really touch these until I was in grade school and even then only for about ten minutes a week during "computer time" on the old monochrome text display screens. Now decades later and after having spent much of my working life sitting at one the information flows from me to the computer by these instruments without my even consciously thinking about it most of the time. It takes almost as much effort for me to arrange my sentences in speaking to other humans as is does for me to use a keyboard and mouse. Coming generations that will soon be running these plants will have users who will have grown up with a keyboard and mouse and it will not be the cumbersome tools perceived by many people that are experienced in the field now. Rather just a functional extension of our hands.

Think of an automobile's steering wheel, pedals and shifter. They could certainly be made "easier" to use or at least apparently easier to use. The technology exists for a driver to just sit there and look at a turn or even just think about it and the car could steer itself there. More practically would be just holding onto a transmitter and tilting side to side, front to back or lots of other movements like playing a Wii. Certainly that would be less complicated than our old fashioned steering wheels, pedals, shifters and all that, right? Because these interfaces are familiar to us and we have grown so accustomed to them anything else would be difficult and just plain dangerous to use until it became second nature as our standard automobile primary control mechanisms are now.

Now say a Formula One driver is told that for the next race all they have to do is hold this little thing and manipulate it to control the car. Simple; just tilt it or twist it to tell the car what to do. It sounds so easy but you can bet that driver would come in last if even finishing the race. Even after a little familiarity and practice I have a hard time believing it would compete with a traditional driving interface.

In short, I know people rave about gesture controls and I'm not discounting the technology. In SCADA HMI's we must think about it before we just throw in the newest technology though. Of course its creators and proponents will push it and tout its superiority just like any product marketing but that doesn't mean it should be accepted for every application it is available for.

Common keyboard and mouse it is for now, at least in my experience and opinion. Please feel free to write me with any real world installation of other control methods. I would be excited to read the results of the implementation and if the findings are beneficial, I may revise this section for future printings with more options and feedback.

Back to communication now. Don't worry, we will not be getting into the specific languages that are used in computer programming. That's an entire science that I'm glad there are other qualified professionals working in. Those languages evolve so fast that by the time you read this book they could be obsolete or at least portions of them. There is programming involved, but primarily we just use a mouse and keyboard to communicate with the computer.

A working relationship between a human and a machine is like a relationship between people. Effective two way communication is the key to all relationships between people. Likewise, the computers must not only receive instructions from humans, it must communicate back with humans. Clearly it can't talk to us directly. It must use other means to translate it's needs, intentions and questions to its users. Mostly this is done by way of a monitor that displays pictures, data and messages from the machine. This display is commonly referred to as a Graphical User Interface or GUI. Often this is lumped together with the inputs and called a Human Machine Interface or HMI.

HMI design is a far reaching term that refers to

any time a human and machine must work together. Examples of interactions by way of HMI's include using an ATM, driving a car, using your mobile phone, using your home computer or even using a lawnmower. While it does apply to these and other applications as well; it is also the primary term used to describe the interface between the operator and the DCS or SCADA system; although you may also see it referred to as a MMI, HCI or others listed in the glossary.

Designing these interface graphics is the real focus of this book. They are how the machine communicates with the human. This is why we spent so much time so far discussing some of the capacities, strengths and weaknesses of the humans and the machines.

We already know that a picture is worth a thousand words but what we forget sometimes is that those pictures need to be speaking a thousand words in the right language. Maybe sometimes that is to much, we may only want to speak five words instead. Pictures, like words can ramble and be filled with fluff. Similar to an entertainment news program that makes you have to dig through it to find anything remotely informative or news worthy.

Pictures can be redundant; stealing your attention long after the useful information was conveyed, or obscuring it. Pictures can be confusing like trying to find Waldo. What you're looking for may not be particularly hard to find nor is the picture inherently tricky. It's just that the task of having to look at hundreds of the wrong thing that are similar can make it difficult to find what you are looking for. The similarities make it take more time to analyze. It's easy for a computer to check differences. It has specialized shortcuts just for that. We don't always have the same shortcuts in our minds.

Pictures are great because they can show a lot of information all at once instead of sequentially adding data to a train of thought like reading text. Sequential data transmission is the way computers think, but we have designed them to turn that into pictures to make it faster for us to get the information we need. This can allow whatever information is needed to be found quickly. Similar to looking at a map. Reading a map in text would be confusing. As a map, however, it can be very useful. Maps are great because you can put loads of data in a small area and it can be very useful and understood quickly in our human minds by drawing on our visual memory and organization methods.

Pictures on HMI screens for monitoring SCADA and plant operations are not static. They are dynamic and dynamic pictures are generally referred to as graphics. Graphics have data values that are changing, equipment diagrams that sometimes change with state and conditional changes and so on. Graphics are more like video games in that they change in reaction to what inputs are given to them. Unlike video games, however, the equipment represented is real with very real actions controlled by it. Again, the purpose of the interface is not entertainment, I can't say that enough. In fact if it is too engaging and difficult, like a video game, valuable information can be missed or misinterpreted. Misunderstanding an HMI can have serious consequences. That said, video games are still probably the closest analogy to the function of graphics.

Just as we have alternative input mechanisms, like gesture controls, there are alternative communications for HMI faces also.

Translucent screens and holograms are the most

popular. Practical holograms are still a long ways off, maybe they'll have application eventually, but they are not a viable option right now. 3D monitors are feasible also but don't have high levels of software available yet. I have yet to see true 3D software for rendering of process control graphics. I'm sure they are coming and I can't honestly say it's a terrible idea but it should be thoughtfully implemented when they do come, not rushed out as soon as it's available. For now, true 3D interfaces for SCADA remain undeveloped, so they are not a viable option yet either. Even if they do come, most of the topics covered here will still be applicable. Maybe in future editions I will be able to address these after we have something to actually kick around and test.

For now, traditional tube displays and flat panels are the most common. Obviously tube monitors are on the way out. They are heavy, energy intensive and just plain bulky. A little more subtle reason is that we don't feel high tech using them and consequently we subconsciously engage with them less. Illogical, but it's the way we are. LCD panels are probably the best current option and the most widely implemented today. They are relatively cheap, have high resolutions, high refresh rates and come in many sizes and aspect ratios. This is a field of high technology turnover and progression and this paragraph can probably be ignored, you now what's out there now and what will work for these graphics.

I have to cover one more thing though. Lately, Hollywood has been in love with translucent monitors. They are in everything from *Minority Report* to *Avatar* and even into comedies like *Date Night*. It's always with the assumption that this is the future of tech; that high tech systems will have these. I'm not going to lie to you here either. The geek in me wants one too. It doesn't get too much cooler than that. The soft blue glow from the back lighting to the sleekness of the display that looks like a sheet of glass with a thin surround. Even better is knowing this tech is completely available. In an environment where undue distraction should be avoided at almost all costs; being able to see through them to the other side is just one distraction after another waiting to happen.

Do you remember the way movies used to have people drawing on glass with dry erase markers so it could be seen from both sides? It wasn't that long ago.

Low tech but it kind of looked cool at the time for central control headquarters. Notice that today old school white boards are still being used despite the fact that a pane of glass wouldn't really cost much more. It's the same idea with translucent LCD panels, they're just not ideal for information communication no matter how cool they look.

We've covered the HMI hardware you'll probably use, or at least some of the considerations to think about when choosing them. That's just the physical face of the machine. Like a blank human face. Without expressions, nothing has been communicated. It is hard to read someone's face who isn't making any expressions. In the coming chapters we'll get into giving the machine expressions for it's face, instead of just teaching it speech structure, or establishing it's features.

Chapter 7
Communicating with Humans

Computers are well suited to gather, store and process large amounts of data. The data they store is in ons or offs; signified by ones and zeros. This is commonly referred to as binary. This simple function allows it to be processed at a very high rate by digital processors. The computer gathers this data, processes it and then controls outputs or stores it for future processing. Maybe a millisecond later or maybe a year later. That's a computer at it's core; a complex database of ones and zeros that process sequentially.

Combinations of these binary codes create basic languages, like our letters create our words. The computer can then use these languages to calculate and store huge amounts of information mind bogglingly fast. I'm not sure we're readily capable of even understanding the speed these are carried out at. Humans on the other hand may not process raw calculations anywhere nearly as fast, but we can often process targeted information faster without digging through warehouses worth of data to find what we're look for.

There is a core difference between how computers think and how we think. Both are capable of interpreting each others but some translations are required. Just like when two humans work together and speak different languages the efficiency of the translation directly impacts the efficiency of the work the two are doing together. Luckily humans can speak different languages but we also share another language; one mostly interpreted by our intuitive mind and common understandings.

This somewhat universal language consists of many subtleties like body language, facial gestures, hand gestures, vocal tone, speed and body positioning. We may speak completely different structured languages but many of our mannerisms are universal. This aids us in working more efficiently with coworkers who speak a different verbal language and even with those who speak the same.

Think of basketball teammates that work together. They can look at each other and know exactly what the other is going to do without any structured communication. Sometimes a snapshot of someone's body language can articulate more than they could if they sat down and explained their intentions and intended actions. This language is often more capable of communicating ideas and interactions in real time; rather than discussing them and deciding a course of action. Life is dynamic. It doesn't happen in a set of sequential calculated decisions like a chess game. Variables are changing all the time and even the change in time ads an additional variable. Processing communication in a dynamic fashion is essential to teamwork in real time.

This real time teamwork is what is going on between operators and automation systems. The more we increase the efficiency of this communication the more we can have people and machines work together with their respective jobs and strengths. The same way a good basketball team can work together with their respective strengths and roles.

People work together with computers every day. We are able to translate each others languages. Many people even know a computer language or two with some efficiency, but still significant translation is required in both directions.

Translation from human language to computer language is making great strides as we've gone over already in the previous chapters. People are trained to talk to computers. It's almost second nature to many of us now. It is like living in another country where it is not our first language but we are fluent enough to function without putting much thought into how we communicate with the foreign language.

The mouse, keyboard and associated software do ninety percent of the translation work; the other ten percent is just us figuring out what we want to say to the computer. That's straightforward and there are books, online resources and academic classes in learning how to communicate effectively with computers and learning all the various languages.

A computer being able to communicate effectively with humans is a different story. Often efficiency is ignored in favor of aesthetics. Dynamic computer monitor graphics are a great start already. They are a huge leap ahead of semi-static serial graphics that preceded them and still exist in many plants. Computer terminals with Graphical User Interfaces (GUI) started gaining traction decades ago for automation control. Recently the technology has been developing more rapidly. Primarily as a by-product of the technological advancements in the entertainment industries of film and video games that are then adapted for industrial control use.

SCADA GUI design has been evolving based on the goal of making them as realistic as possible. This influence is often adopted along with technology when being adapted from other industries. Computer animated movies and video games are pioneering the actual graphic display capabilities generally. This idea is an obvious inheritance from their donor industries and not usually questioned because it is interesting and we don't usually question things if we like how they are. I can't really blame anyone that would prefer to work with realistic rendering for automation graphics. They are so much more fun to work on. It's interesting and you get a real sense of accomplishment when you design something or animate something in 3D that mimics the way we see the real world.

Entertainment is not the goal for interface design

though.   Websites are somewhere in the middle of aesthetics and function.   User engagement is the goal; getting users to either buy products or to stay around as long as possible.   There is a lot of money to be made with websites so there is a lot of research done in the areas of human factors and interactive psychology. While the research is applicable the application is very different.   The ideal website shares some similarities with GUI's for SCADA but has some very different goals and thus should have very different approaches. In the coming chapters you'll we will look to websites often not as models for graphics but recognizing elements that exist in both.

Chapter 8
Teaching Computers to Talk to Humans

We could go straight into some of the basic design mechanics. Instead, to get us into the right train of thought, let's imagine an example.

Think of the classic arcade game where you have to shoot bad guys that pop up, but not shoot the civilians that pop up. That is essentially what operating a plant by reaction is. This can be difficult when the picture is either too vague, has incorrect context, or is too realistic. The realism is what makes the arcade game fun, but that fun is generated from the difficulty of perception.

Now let's put on our designers hat. How can we take that task and make it easier to accomplish without just simplifying it and making it slower or having fewer bad guys and so on. How can we keep the same speed, the same target area and the same variables but make it easier to accomplish the task accurately and efficiently. Clearly it would be less fun but let's see if we can make the task much easier to accomplish accurately and quickly.

First we could change the bulk of it to gray scale. Make all the objects in gray with a darker gray or black object outlining. Next make the background a light neutral gray. While we're at it lets take out everything unnecessary; all the embellishments on the objects, all the extra little details.

The important part is the active objects, the people that pop up that require action be taken. We can start by using colors to designate action. Not like green for safe or blue for enemy. Let's think about the actual task again. We want to click on the bad guys but not the good. There are people popping up or moving around on the screen all over and as long as they are not supposed to get a reaction out of us we have to see them there as per the outline of our proposition. It must have the exact same variables, speed and so on.

All the people will still be popping up but they do not all require your action. The civilians need to be different than the background and other contextual objects but different from the bad guys. Let's make these other people that are there a charcoal gray. So we see them and they are popping up but we are not really visually stimulated by it. We see it but it doesn't get our heart jumping and we can be looking next to it but our attention is not grabbed by it.

Now for the bad guys. We want to see them and the faster the better. We have to instantly recognize them as different from everything else on the screen. We need to know as fast as possible when something needs our attention. Let's try color coding them red. Sort of an international color code for urgency.

Now we're watching this city street scene all in gray scale. Things are moving around; some things in the foreground, some things in the background, but all in gray scale. Then a bad guy pops up that is red. The first time it happens, it sort of surprises you, but you click on it. Then the next one a second later. No problem, you got him. Then the little old lady in dark gray walks by. You see her but you're reaction isn't first to click on her. Then on the other side of the screen a red bad guy pops up for a second then, Bam! you got him.

It wouldn't be long before your reactions started

playing the game for you. You could probably even play the game like this while carrying on a conversation with someone; something not likely while playing the regular game.

Maybe I should steer clear of arcade game design as I think I successfully took the fun right out of this one. Fortunately we're in the field of trying to optimize the efficiency of a task; make it easy even. The example was easier but not because we slowed it down and not because we had fewer distractions or anything like that. Just by changing the colors we increased the efficiency of the task. Now we can be done with ruining the arcade games.

People often understand the importance of standards for the use of colors. It seems like a given but it is almost always done either arbitrarily or maybe it is voted on; but often it is just loosely based on traffic lights or something someone saw somewhere else and often on vendor templates.

This was one example of how changes in standards can change the efficiency of the GUI. There are many other factors we can work with as well. Factors that are already in place but often not considered in designing graphics.

We've already walked through an example of what some basic color changes can do. In the rest of the book we will briefly talk about the effect of color on our communications as well as other factors and concepts we can use to our advantage and some just to be aware of. We will not go into specific symbols to use or try to build your templates or color standards specifically for you. I want to give you the basic tools to do this for yourself. Up to this point we have been establishing the need, the cause, the condition, the languages of people and machines and the roles of the two.

In the coming chapters, we'll go over a few of the basic human shortcuts that can be used by computers to talk to us. These are just a few of the major shortcuts that use our intuition and built in cognitive reflexes to our advantage. We like to use shortcuts in the computer to communicate to it the things we want it to much quicker than by manually typing in a complex string of commands. The same can be done with humans, if we know the shortcuts are there and take a little time to consider them while designing graphics.

Applications

Colors and Audible Alarms

Polarities and Black and White

Context and Static Content

Natural Eye Path

Are Left and Right Symmetrical?

Alignment and Scanning

Trends, Bar Graphs and
Small Multiples

Mental Navigation Maps

Chunking and Working Memory

Chapter 9
Colors and Audible Alarms

Color draws attention that's why we love it. Humans love color. We love it in art and we appreciate it in nature. We feel connections to different colors and the way it makes us feel and think. We use colors to describe our many human emotions and thoughts. Why do we pick colors for our vehicles, our phones, our walls and everything else we have a choice in? Wouldn't it generally be more logical and efficient to just have each product made in a single color; whichever color is cheapest to produce?

Colors no doubt contain complex human intuitive connections. This is why colors can be valuable in designing graphics. If they are merely used arbitrarily to build some standards and templates we are using the strength of color in a counterproductive way.

Again; color draws attention. As a general rule of thumb we only want to use color when it requires the attention of the operator to be grabbed immediately. Some feedback from operations about primarily gray scale graphics is that they are "boring and dull". That's usually a good sign that they may be a good set of graphics. If an operator's eyes are glued to one screen because it's colorful it will not be drawn over to another when it's needed.

If you come from a design background you probably don't like the idea of continually putting out work that is designed for people in general not to enjoy looking at. It doesn't give us that warm fuzzy feeling. The satisfaction of people telling you they like or appreciate you're work. Maslow tells us that desire is completely normal. Just because people may not appreciate your graphics doesn't mean you didn't do a good job or shouldn't be proud of your work. I just want to prepare you for the fact that people will probably not be impressed at first with your work if you designed the graphics well, they're designed not to stimulate your mind with static context.

Use colors intentionally, strategically and sparingly. Colors talk directly to our "right" brain first. Remember the shoot 'em up arcade game? A sparingly but strategically used color can increase reaction time and accuracy.

Often users of my graphics get the impression that I think colors are bad because many of my graphics may have no color on them at all. That couldn't be further from the truth. Color is a powerful tool but it can be overused.

A typical color standard often used for interface graphics is the generations old red for stopped and green for running. I make a rash statement now that I'm sure may not always be the case. As a general rule of thumb never use color to show regular or intended operation of equipment that requires no intervention from operations and will continue running fine if nobody saw it. Colors should be used to signify that operators need to intervene. This can be either for predictable and intended interventions like to go take a sample from a tank. Colors can also be used for abnormal situations that need to be addressed right away.

Probably the most important color suggestion is for the use of red. You probably guessed that. Using red for something being stopped, shut or off is an extreme example of crying wolf to your right brain.

Eventually in that context you will start to ignore your reaction to take notice despite your inherent reactions. A better use of red is for displaying something that needs immediate attention. I'm sure there are many books written and many people who have done a lot of study on why red has an intuitive connection in us with needing attention. Maybe it's just because we were always told to pay attention to things in red. It's clear that red can speak to our intuition and get a reaction stronger than other colors. The computer doesn't know that intuitively but you can teach the computer to use that human shortcut to more effectively communicate with us.

Another basic standard I often use is yellow or orange to signify when intervention is required but is expected. For instance when a truck needs to be hooked up or unhooked. It is not an abnormal situation. It is expected and controlled by the SCADA but it does require a human to do something or make a decision.

Yellow and orange typically trigger more energetic or action based concepts not reaction based concepts. You see a yellow light and you take notice but your gut instinct doesn't tell you it's an emergency.

Rather something that you do need to see and think about.

You may or may not want to use an audible sound for these instances. That is up to your plant, its complexity and the platform capabilities. An audible indicator like a ding or something can be a good addition to help operations with performing tasks. Usually indicated in yellows or oranges on the graphics. Likewise an audible buzzer or alarm is likely already in place for instances where red would be animated on the graphic.

These audible alarms are probably already in place but coordinating the audible and visual alarms is helpful in efficient abnormal situation management. Audible indicators can signal operations that there is something needing attention somewhere. Hopefully there is a means in place to help locate the source. Most platforms have some sort of alarm management system built in. If not you can build one graphically showing all the alarms in the plant with navigation to its unit graphic. When the user hears an audible alarm and they are unsure of its source they can go straight to that sheet, see the alarm and go straight to its unit sheet and see what needs to be done.

Something else to keep in mind when thinking about color and audible alarms is that there are likely at least a few individuals in the plant that have some level of color blindness that can inhibit the recognition of visual alarms. The audible alerts can supplement this lack of visual recognition. Maybe you are not color blind and you don't think anyone in your plant is but there is a good possibility that someone is. While only half a percent of women have some form of colorblindness, roughly eight percent of men have some form of color blindness. This is a significant enough number that it should certainly be considered in designing the computer to communicate with humans.

Blues and greens generally trigger a sense of calm. For this reason I use it for other non-imminent signage occasionally when necessary. One plant I was working on had several nearly identical reactor trains. Before I was working on that plant there was an incident where an operator thought he was on one train but really had the graphic up for another. He proceeded to execute a command based on the information from the wrong train. This was caused because the graphics also looked very similar. Nothing jumped out at the user that it was different. The title at the top was different text but it didn't jump out at him. It was decided that this could be an exception for the use of color and people before me with good interface design practices decided to use colors on the title bars to designate which train the unit was on. After that the titles blocks of all the equipment for the different trains were highlighted in a different color and I don't believe this kind of mistake was made at that plant again.

Using reds and yellows should be avoided in the case of signage for static content. When using color standards avoid using the same color for different priorities as it can be like crying wolf and detracts the power of those colors from their primary purpose in that set of graphics. Different shades of blues and greens are great for static content but always keep the graphic titles in exactly the same place so people's minds come to expect it there and they don't spend time searching for it. It will become reaction.

I'm not sure I even need to go over it at this point, but I have seen it on many different platforms and packages. For goodness sakes do not use colors for backgrounds or static equipment. We'll get into that in coming chapters as well but it's applicable while we're discussing color usage. Of course as soon as I throw out an absolute I'm sure an exception will come up but its still worth throwing out there in the mean time.

Use the most neutral colors possible if you really want to use color for a background or object. Light tan or brown can make good backgrounds but still you see it and you never want to see a background. Maybe you are designing an HMI for something other than a control room or remote process control terminal. Or you just really want to use a color for the background. If that's the case pick a pale background preferably in the tan or light brown family. Avoid reds and yellows as those are likely being used for other important notifications. Your eyes can't help but look at yellow and red. You'll keep drifting to it and scanning it but we never need to do that for a background or static equipment. You might not even notice it with a background but it will be pulling your subconscious attention.

Chapter 10
Polarities and Black and White

We use gray scale to represent polarities because that's how we think about gray scale; as polar ideas. Think about when we consider the extreme polar concept of Good and Evil. We describe the polar concepts by using black and white. When we think of something being a portion of one and the other it becomes shades of gray but still the intuitive understanding of the two components is in black and white. The blending of the two is still based on the polar concepts and the portions of each opposing ideas.

We can use this. Our minds already subconsciously think about things as polarities. We have a hard time sometimes thinking of things otherwise. Computers understand polarity well. That's how they think at their very core. That's why they can process data so efficiently; it's easier.

The logical portion of our brain knows that much of the world is analog. We see the added benefit we sometimes get by using analog spectrum ideas like color because we have so many more options than just on or off. When humans get involved we know that the world ultimately is analog. Things are not on or off generally. They are not either light or heavy they are just what ever they are. Without the concept of polarity neither have any meaning though. Most of these usually come from a combination of polar ideas however and it requires our valuable thought process to analyze the ratio of the polar ideas for a particular value. We're used to it so we don't consider the amount of thought that has to be spent but we can be more efficient if we don't force ourselves to spend that extra thought to analyze the ratio.

This whole idea seems like a moot concept for interface design. I assure you though this has extremely useful application. We as humans think in polarities or ratios of polarities. Conveniently much of the equipment we are controlling and monitoring is also in polar states that we need to know. Is the valve open or shut? Is the pump running or off? Is the light on or off? This is something anyone in process control understands.

These digital inputs (DI) and digital outputs (DO) are the most common variables we have. Even analog variables are really polar ideas. The analog valve is not fully open or closed but is some portion of both. We still generally look at partial values as portions of the two polar ideas.

Let's step back and look at the way our minds work in this regard. We often subconsciously categorize things in polar states requiring less memory and processing. We inherently think of dark and light as polar ideas as well. The situations we want the machine to communicate to the humans are usually in polar ideas also.

We look at these two basic ideas and it's more obvious. We should program the computer to show these polar variables using black and whites. Deep jet black and bright stark white often attract attention and we want to merely convey information without attracting attention most of the time.

One way to use our mind's polarization without attracting attention is to use a dark charcoal gray and a very light off white. The kind of off white that seems to attract no attention at all. Our first reaction is that this bland combination is a bad idea. It seems too neutral, but it works with our subconscious very well. When you see a graphic with sixty digital variables that use colors like red and green to signify the state it can require much more mental processing power and risks your brain being distracted from things it should be doing. You can look at the same graphic using dark and light to signify the states and you don't even see it but you still know what is going on with the unit.

Even more importantly you see that red circle next to the level indicator telling you the level is higher than it should be or you see that level shooting towards one of its parameters. That is the real beauty of using our subconscious polarization to convey the information that we do not need our attention. The real point is that all of this other information is stuff we need to be able to see but only the things that require our attention do we see first because they attract our attention.

This is a concept the computer doesn't inherently understand. It's a concept designers often overlook as we think in terms either of graphic design, schematics or computer languages. It looks bland when we look at interfaces in the light of a display that is supposed to captivate our attention. When we think of an interface as a display to communicate the information we need when we need it these bland graphics can do the trick. It is the key to prioritizing the relevant information.

Chapter 11
Context and Static Content

Context is one of the biggest assets a graphic has. Without context data is just values or states. Context displays quickly and intuitively how all the data relates to itself and the equipment it is monitoring and controlling. The bulk of the context is the equipment being controlled. This is an aspect of interface graphics that requires our attention. It typically is not dynamic. It's just the pictorial representation of the equipment. Usually with labeling and very roughly scaled to the field.

Attention must be paid to designing static graphic elements. They are important but that doesn't mean they should be the most prominent features of the graphics despite the fact that the equipment they represent is physically the biggest in the field. They should convey information almost exclusively to your subconscious. The user shouldn't think about those elements any longer than necessary as they open the graphic or walk up to it. They should think about it just enough to know what they're looking at and after that they should see the variable information that tells them what is going on with the graphic they're seeing. Is there material going in or out? Is the agitator, blower or pump running? These are the things the user is supposed to understand as soon as possible so they can perform the task they need to or just confirm it is running properly and move on to the next unit.

Color attracts attention even long after it is initially processed. Your eyes will keep jumping back to it. We do not need to continually re-establish the context of the information on the screen. For that reason using colors, particularly bolder colors, should generally be avoided for backgrounds as well as equipment elements. Similar to backgrounds; if colors are required for contextual elements it is preferred to use muted tones from the blue, green, or brown palettes. Avoid yellow and red still even in muted tones. These color suggestions will lend attention primarily to establishing an understanding of the information that is being displayed instead of just looking at all the equipment.

Gradients are another popular tool for interface

graphics. These are often used to simulate light reflection around a cylindrical, conical or otherwise rounded surface. This does make it look more realistic but it is not necessary. Whether it is recognized or not gradients require mental processing to interpret. The half a second or so of processing power may not seem important but for someone whose eyes may pass over that equipment hundreds of times a day it can add up. It's generally best practice to avoid unnecessary mental processing unless it is used to increase accuracy, efficiency or safety. If a good case can be made for why gradient is used in a particular situation then certainly go ahead and use it. Be aware though that it can cause a conflict of attention; especially when used in conjunction with embedded trends and graphs that we'll cover later.

Shadows are another popular tool that often gets used in creating graphics. Maybe they can be useful in some situations but primarily it is embellishment to make it look more realistic and for no other purpose.

This is a general principle for design but especially for context and equipment elements. It is generally recognized that only around forty percent of a screen's total area should contain its element contents and sixty percent should be background. This is counter-intuitive but often not saying or showing something can lend it's portion of attention to the things that are chosen to be shown. We should avoid the urge to add things that don't need to be there. If it adds nothing but takes away from things that do need to be there we must refrain.

Analyze the human factors that are necessary when viewing a particular graphic; particularly attention and mental processing. Then analyze the potential benefit from adding additional touches like gradients, shadowing and other objects to a graphic. Then decide if there is an additional safety or efficiency that is gained by having those touches. If there is a benefit to having something and it outweighs the drawbacks of added visual complexity and stimulation then by all means proceed.

Backgrounds can be a place for attention to be attracted to as well. It might seem like it wouldn't matter because it's not showing anything and you'd think people would just get used to it. Having attention grabbing colors for a background is a common mistake. Because while the users will get used to it, it will still distract the eyes and attention. Black is a common background as well. Left over from early computer systems. Black backgrounds are on the right track as the intention usually is to be a background that doesn't attract attention. Black is still a large visual weight on the interface though and is not as effective as a neutral light gray or off white. The ultimate goal with background color selection is for the user to not even be able to see it. Black and colored backgrounds however are able to be seen whereas the neutral lighter gray is something our minds do no really register as anything. Not even as a big blank colored screen. The goal of the background is to attract no attention and this directly lends more focus on the actual information on the graphic.

If the sole purpose of an element is just to make the picture look more interesting it is better to pass on those no matter how much you might want to add them.

Chapter 12
Natural Eye Path

Eye path is a short topic but one worth noting and considering during user interface graphics design. We humans generally look at a page or computer screen and look in the same locations in the same order almost every time regardless of aspect ratio. How we arrived here could certainly be debated. Nevertheless we have a basic eye scan path programmed by the time we reach adulthood. This shortcut may be a tool or may just be a handicap. Regardless of how it ended up there it's there in most humans and should be acknowledged.

We typically look at a page and quickly scan it in a predictable path looking for the information we went there for. We start at the top left then scan over to the middle of the screen. We then look to the left side about half way down and then do a quick swoop along the bottom. If we don't find what we're looking for by then we begin getting slightly frustrated and start thinking

119

about it more and searching for what we are looking for. If we are sitting in a cubicle designing graphics we may not be so quick to get frustrated at not finding what we're looking for right away. When you are running a plant or using an HMI for some other useful application you want the information you went there for and if you don't find it in the initial natural eye path scan that's when the frustration process first begins.

We can not typically cram every useful thing on a graphic into the view of the natural path and we probably don't want to as we don't necessarily want everything to be processed in the first scan. Being aware of natural eye path can be a useful tool in effective design when considering the information order you want to be processed. Maybe try to start with the name of the graphic in the top left so our mind will know what it's looking at. Then shoot for having the main unit on the graphic being roughly smack in the middle of the screen. Next put all the feeds or inputs on the left side of the screen almost like a menu bar at the left side of a web page and finally have the outlet on the bottom right.

I realize that I just made some pretty rigid standards recommendations there. And feel free to use them or not; but you'll probably find this is a fairly standard practice even though many do not know why. The reason we do it this way is because it follows our natural eye path movements and leads to the quickest way for the user to understand what is going on with the unit. This results in the least amount of frustration from cognitive processing energy. That setup along with the other practices of effective interface design can allow users to open a graphic and within a few seconds know what is going on with the unit. Right away they can know where everything is at and move on to the next unit or graphic.

Again this is a small section but one definitely worth covering. There has been a lot of research done on the topic. Most of the research is primarily done in relation to website design. Notice any well designed websites typically have the title in the top left. Typically with the navigation location right below that lets you know what site you're at and where at that site you are. Then the prime attention elements are in the center about two thirds up. On the left hand side you see the menu with major site categories then down to the bottom right portion of the screen typically is the control for scrolling down to additional content within that category.

Some things can be gleaned from mainstream design. Websites are somewhere between utility and entertainment. They want you to stay at their site but they know they have to convey the information you want in the order you need so you don't get frustrated and leave. This is something we can learn from all the research and development that has been conducted on website design.

Chapter 13
Are Left and Right Symmetrical?

This section is about a topic that is a little more controversial. Not about it being one way or the other but rather in whether the pattern really exists or at least the extent of it. Nevertheless from observation, research and practice I have come to a general understanding but I look forward to learning more in the future. Or maybe it's as basic as my understanding and that's all there is to it.

The topic is our field of vision and brain function. Primarily left and right fields of vision and left or right brain function. The answer probably lies somewhere in the study of the extent of the right and left brains' functions. Our application is particularly in regards to visual comprehension and subconscious preference.

Do you have your own workspace at a desk? If so did you have any control over the layout? If you answered no you can skip the next few paragraphs or take some time to walk around and look at other people's work spaces that were able to lay it out themselves.

If you do have your own I'd like you to step back for a minute. Look at your workspace. Primarily look at whether different elements in your workspace are logical elements or intuitive elements? Are there schematics, blueprints, drawing markups, billing statements, contracts, technical computer applications or other logical elements? Also are there pictures of your family, your car, your house or your hobby? Maybe a motivational or humorous calendar or maybe desk plants, a fancy paperweight or any other item on your desk that isn't really a logical element. Mentally separate the elements into one of those two categories if possible.

Look at where on the workspace those are located. Are the majority of the logical elements located on the left side of your workspace? Are the majority of the personal or intuitive elements on the right side?

Yours may not be but most are. Sometimes there are logistical reasons for not having it this way. Where the options for your computer are, where windows or doors are and so on can affect this also. When possible this is generally the way we chose to arrange our environments though. Whether left or right handed, creative or logical personality types; it doesn't seem to matter. Most people lay their workspace out like this.

Why? Since most of us are right handed wouldn't it make sense that most of us would put our technical stuff on the right side like blueprints and drawings to mark up? I would think so. Maybe you do; but when all other factors are the same and people move into their office or cubicle and start unpacking boxes it usually follows this layout.

The first day I thought about this and read some research on the topic of field of vision and brain functions I started walking around snooping around my coworkers' offices. More than three quarters of them had a similar layout of logical elements on the left and intuitive or emotional elements on the right side of their work spaces. That got me thinking. Are there any applications for this in my work?

I started looking at websites and I noticed this organization at work as well. As we already covered in the previous chapter most of the highly effective websites have their title in the top left and their category menus on the left. Those are both logical elements. Where are you and what are your options here? These are information elements that are processed by the logical portion of your brain. We also pointed out that the most valuable content is smack in the middle. This is information that borders between logical and intuitive. If it's a blog, an article or something where language is the content then it is generally in the center. Language is both logical and intuitive. It is processed by both portions of our brains. So being in the middle is the best way to convey information to both polarities of our cognition.

The portion of most websites we haven't covered is what is going on at the right side of the site. Typically on the right side there are ads, product recommendations, links to other sites and other emotional or intuitively charged elements. Take a look around if you're skeptical. There are other options but it is the most common standard layout for high functioning websites.

The application in process control interfaces is

that it is yet another human shortcut we can utilize to aid in designing graphics to use existing human cognitive processes to help us design the most effective graphics possible.

By arranging, when possible, the logistical elements on the left side and the intuitive elements on the right we can increase situational awareness of the user much faster and with less frustration than when we fight it. Less frustration means greater focus and accuracy and more productive users. Logical elements are things such as unit phase states and navigation elements while intuitive elements are things like embedded trends and alert panels. One basic way to tell the difference is whether it is an element that you might go looking for or whether it is something that you should just see. There is a great deal to be learned in this field and I look forward to learning more about it myself in my continued studies.

One more element on the topic that may have application in situations I haven't yet looked into is the connection of the optic nerve. It is important to note that the optic nerve from both eyes goes to both sides of the physical brain but the left field of vision of each eye goes primarily to the left brain and the right to the right. It is not that the left eye is attached to the left side of the brain only and the right to the right side only. This is just something to remember in regards to visual layout and associated thought processes.

Chapter 14
Alignment and Scanning

Aligning information allows us to scan very fast. We can accurately get an idea of the condition of a lot of variables at once. Some graphics may have twenty or more materials going into a vessel. If the valves on these materials are scattered all over the graphic as they often are on P&IDs we can still find all the information but it can be a real chore. Trust me; it can be frustrating when you are trying to figure out what is going on.

By arranging all the inlet valves in a vertical line, using the dark for open and light for closed standards in a very short time we can have an understanding of what materials are going into the vessel. Likewise if you have that many outlets going out to storage tanks or something you can quickly know where the material is going.

This idea can be used also for arranging columns of flow rates, pressures and so on; especially if you are able to add bar or line graphs. If you have twenty analog values scattered all over the place you can look them up but it can really chew up valuable mental processing power. Think now if you line up all the variable values. You would see much faster what is going and at what rate.

We'll look at bar graphs and small multiples in the next section but I can not talk about alignment and scanning and not mention them. Think again about the set of twenty flow rates. Having the values lined up is a good start but now think about having bar graphs of those values all lined up. Now how fast can you scan it? You could have fifty horizontal bar graphs lined up vertically and you would see right away which ones are at full flow, which one's are stopped and how all the flows relate to each other. The same can be done with pressures, percents of completion, power to various motors and all kinds of different information. Get creative! This is an often underutilized but very useful and basic concept.

Chapter 15
Trends, Bar Graphs and Small Multiples

The idea of small multiples usually applies to a set of mini trend graphics. Dr. Edward Tufte is a pioneer in information display and popularized the concept of small multiples several decades ago. I'm sure there were some before him; but he explored, organized and articulated the ideas very well. If you're interested in understanding this topic in greater depth you should definitely look up his work. The basic idea is that by having very simple, non-embellished multiple variables displayed with parameter relevant context one can see a lot of information in a small area and very quickly. For an example pick up any financial newspaper with recent stock trends listed. They typically have a whole row of trends for different stocks all with a relevant scale and all with the same time span.

We described the advantages of bar graph usage

in the previous section on scanning and alignment but there are some more aspects that should be covered. The principles of color apply to embedded bar and line graphs the same as it does to the rest of the graphic. Colors should be reserved for things that need attention. The tendency is to use color when we have the option to but it should be avoided unless necessary.

Occasionally colors may be useful for trends with multiple variables being tracked on the same line graph. If colored trends are needed try to use muted color tones. Sometimes the graph is intended to be the focus of the graphic and all the unit information is just to give a broad picture of what is going on. This doesn't come up often but occasionally it does. Colors could be used in that situation a little more liberally since these graphics are not typically used for plant control but more often for plant monitoring and study. For those where efficiency can be sacrificed for abstract study of overall plant or unit operations they can be approached differently maybe. At any rate after reading this you should be able to make that judgment call.

Trends can be very useful in controlling plants. People sometimes think trends are just for monitoring and studying. Something to look at to get a picture of past performance. What is often missed is that trends offer valuable information about the dynamics of a unit operation. If a user walks up to a graphic or navigates to it, it might be unclear from the static and variable information what is going on. You see the level but is it rising or lowering? You see the pressure but is it increasing or decreasing?

Embedded trends can lend relevant recent history to the variables allowing the user to see not only the current values of the variables in the unit but can lend dynamics to the process. In this situation a snapshot can provide much of the information that would be gained from having been sitting there watching it. This is a leap in user comprehension ahead of current data only graphic displays.

The next step beyond merely having trends is establishing parameter guides. Parameters can allow the user to see if the variable change is heading towards a critical limit. If that variable value is slowing as it is approaching the limit or if it is going to crash right through it we can see that before the limit is breached. At that point an alarm will sound but if it can be seen heading there that alarm can be avoided and there is a much greater chance that a process safety incident can be avoided as well.

Parameters are kind of like lines on a road. You could drive by watching the road rather than listening for gravel to spray up if you're straying off the road or by listening for horns to honk if you're drifting into oncoming traffic. It's the core difference between driving the process and being driven by the process. It's the difference between trying to react to and keep up with the process or working together with the process. This idea is a critical difference between an efficient, safe and profitable plant and one that is having an unnecessarily high rate of process safety incidents causing lost safety and productivity.

Embedded trends particularly with parameter indicators can be one of the greatest assets you can add to your plant graphics. Not every variable needs them and again think of mental processing limits but things like vessel pressures, levels and so on can be very useful in driving a plant unit operation.

One thing that can be even more useful is a target trend line. This is often not something that is available but sometimes variables such as level indicators in closed systems will have an expected level based on how much material is presumed to have been taken out or put in. In cases where we have an expected value calculation available it can be useful to add that to a trend along with the actual value as well. If this can be done we should use a more subtle color tone of the actual measurements color code so as not to draw primary attention to the expectation. Rather to be a guide to make it obvious when the actual measurement is deviating to far from the expected value.

Just the topic of trends could probably warrant an entire book itself, but this should be enough to get you started though. Just remember that a small trend can provide a large amount of useful information and context in a very small space and with low unwanted visual stimuli. You can fit an entire spreadsheet worth of useful information into a small graph. Try to do that with variables alone and it becomes a mess.

Chapter 16
Mental Navigation Maps

This section is one that is almost always visited by interface design books, articles and consultants. I would certainly be leaving out useful knowledge if I didn't at least cover it. This is one area that vendor templates are often fairly good about as well but I want to cover it so you will know if you run into one that might not be.

Everyone knows that any decent SCADA graphics collection needs to have navigation linking it all together into a plant. A hierarchy is nearly always recommended to allow you to quickly navigate accordingly. There are many preferences of how to display your current location, how to display navigation buttons and how to organize them.

You will generally want to start with a plant overview graphic. Usually not a lot of animation here as this is either primarily or exclusively navigation buttons and unit groupings. In a smaller plant you may have every unit operation shown directly on the plant overview. In larger plants this is not always possible. For these mega plants you will want to break units up into operating areas or process categories. You may have navigation to overview sheets that have all the storage tanks on it or all the raw materials or all the auxiliary utilities on it and so on. From there you can break it down and have navigation buttons to all the individual units. This all depends on the sheer number of units your SCADA system oversees.

Computers have no problems just viewing graphics as an alphabetical list of all the unit graphics even if there are hundreds of them. It can just navigate directly to them like navigating to a website using just a URL list. Humans however prefer to have them grouped into manageable sections like file structure trees. We don't mind having intermediate steps to get where we're going if it helps us with our mental picture of the organization.

We assign mental addresses to these units and while a computer can just assign an individual number to all of them maybe up to a thousand graphics in a single plant we can't. Humans prefer an organizational method more like geographic addresses. We could assign an eight or nine digit number to every address location in the United States. Then we could have those all listed numerically. A computer would prefer this. It could find any location it was looking for very quickly and more accurately than using actual addresses.

Humans would take much longer. If we think of those location addresses in terms of hierarchy of categories, usually geographic, we can get there much easier. Think of first listing the state, then the city, then the street and then the number. That's four categories but we can manage finding what we are looking for fairly easily with that hierarchy; even though it requires grouping information with intermediate steps that require extra work and has many logistical data inefficiencies. It just works more efficiently with the way we view the world in our minds. This is probably redundant but worth covering just to establish.

How we indicate the navigational location on the graphic is another point. Typically you have the graphic title on the graphic somewhere. Some platforms have external navigation indicators like a map to the side that shows what section you're in and all that. If those do not exist you will want to have another method.

One way is just to have a small navigation button on every graphic that takes you directly to the plant overview from wherever you are. This is a good start so you can get wherever you need to go somewhat quickly but maybe not optimized. Better yet is listing the location in the tree by having the navigation path by section hierarchy. That way you can click on each sub section that gets you there so you can step up one section or all the way straight to the plant overview.

This is used on many websites particularly when they are selling a lot of items that they can categorize. If possible always have a small section that is the same on every graphic and indicated subtly. This is for reference only and shouldn't draw attention. If you have it right below the title you will get an intuitive mental image of where you're at as you scan past it in your natural eye path. That can help establish what you're looking at also.

Of course whenever possible it's good to have embedded navigation buttons as well. One example is when you have a raw material coming in from the left or a material going out the right; having an actual navigation button right there at the line is very useful.

It's an intuitive place to put it so you don't have to think about where it's at in the navigation tree. You can just go straight there as that would be a natural navigation movement anyway while the task is being performed. Sometimes it will cross to a different place in the tree like clicking on a steam line that takes you to a utility section or something that is far away in the hierarchy tree. This is another case where having the indicators of tree location we just discussed will help in establishing your new location within the plant.

That pretty much sums up a basic look at navigation. There are many ways to show them often just limited by your platforms drawing package. Keep in mind everything else we've covered when establishing navigation standards; things like eye path, color, gray scale, alignment and so on.

Chapter 17
Chunking and Working Memory

Another topic of different opinions is the actual amount of working memory or as it is often called "short term memory" that we humans have. This working memory is the capacity to retain information without storing it in our "long term memory" storage. This allows us to function. It helps us in remembering what we are doing or remembering a short string of number or words.

The cognitive psychologist George A. Miller wrote a book in 1956 entitled "The Magical Number Seven, Plus or Minus Two: Some Limits on Our Capacity for Processing" that used experiments and hypothesis to form the theory that our human minds can only process five to nine variables or "chunks" at a time in our working memory.

More recently noted contemporary psychologist Nelson Cowan has published several works about his findings that the working memory of humans is closer to an average of four than seven and that it isn't necessarily a biological limit imposed on us. Nevertheless it is clear that our human minds are not as adept at working with as many changing variables in the same fashion and accuracy that computers are.

Memory is commonly referred to as "chunks" which can be a single digit, an entire word or a simple image. Think about a string of numbers. How many can you remember if they are read to you just once then you are asked to recall them ten seconds later.

That number can be greater if the numbers are in a sequence; like if they are numerical. If someone read the numbers one through nine to you and then ten seconds later asked you to recall them it would be easy, right? That's because it is really only between one and three chunks of memory not nine as there are numbers of digits. You remember the starting digit, the end digit and that they were sequential. At which point you recall from your long term memory how the sequence goes.

Compare that to the same number of digits but read to you in a random order. Now ten seconds later you are asked to recite them back in that order. That would be pretty difficult, right? Maybe some could do it but most of us could not. That's because that same amount of information is now nine chunks of memory instead of three.

It gets even more leveraged when you think of words. You can see how mental chunking can be utilized and leveraged and why its limitations need to be acknowledged. Each word is generally one chunk of memory and sometimes even strings of words or even whole sentences or more that can be pulled from memory. Compared to the individual characters when scrambled they become individual chunks. If someone read you the following sentence:

"All dogs go to heaven."

It might register as only one or two chunks of memory and you probably could remember it in a string with other quotes as well. If someone read you the following sentence:

"To go dogs all Heaven."

It takes likely a full five chunks of memory at least to remember the string and be able to reiterate it soon thereafter. Now if somebody read you the following sentence:

"All dogs go to Heaven.

145

Now backwards, Heaven to go dogs all."

Then a bit later were asked to recite what was read you probably still could because it is still only taking a few chunks of memory. One for what the phrase from memory was, then another for the concept of backwards and another to remember the order. It only takes maybe three chunks or so to successfully read back over forty characters in order plus spacing and punctuation. Pretty impressive for short term memory.

Now think about this being read:

"To go dogs all heaven
dogs backwards go now dogs all to."

Good luck! Even though it's the same amount of actual content being processes and used in your working memory chunking makes it possible to store much more information.

Likewise images or concepts can be chunks. Anything that already exists in your long term memory as an associated memory can be chunked and utilized in short term memory and processed soon thereafter.

This is the idea of working memory. Now what are the practical limits? Neuropsychologists generally recognize that there isn't necessarily an exact number that is a maximum capacity for humans and there are conditions that can cause people to have much larger or much smaller working memory capacities. The majority of people do use a somewhat predictable practical limitation that we need to be aware of however. I tend to find it is closer to Cowan's estimate of four: give or take.

This understanding of working memory can have a lot of applications on what is expected from users when using a graphic and we need to be aware of user limitations. For example when breaking up overviews try to keep the categories grouped in three to five different sub-categories and so on. When showing multiple units or unit sections on a graphic try to keep them segregated and try to keep the number of groups roughly in that range as well. If we keep them at that grouping size we can more easily monitor the condition of all of them simultaneously. Compared to having to sequentially look through them all. There are many more implications in interface design but just being aware of the concept is the most important idea to take away.

What Now?

You might not recognize what it was you were learning here or if you were learning anything at all. When you design your next set of graphics go back and compare them to previous ones and you'll probably see the difference. My goal isn't to teach you things that strike you as revolutionary or original. Although some of it may seem counter-intuitive the concepts are based on our natural interactions and hopefully it all made sense as you were reading it. With some ideas we may already know the components but reading it assembled and organized can help bring some clarity and confidence to your practice.

I want you to feel ready to tackle the next project with greater understanding, clarity and confidence and most importantly design control interface graphics that make our plants and factories safer and more efficient.

This book was written to empower you to build

your own templates, build your own symbol palettes and dive into the design process. The truth is that in different environments some different standards might be ideal. Certainly having standards across the plant and across multiple plants in a business is generally a good thing.

Occasionally graphics for specialized tasks and situations will arise and don't be afraid to build specialized graphics for those situations. Graphics designed for a board operator that will sit in front of graphics most of the time will require an entirely different approach to standards as compared to designing graphics for a remote GUI with a specific task or a specific situation. One difference is that for a remote interface used by operators that are walking by you may have much larger font sizes for critical values and often show fewer variables but all geared towards performing the task they were created for.

Never be afraid to analyze a situation for human factors and rethink what you know or what you have learned. Even things you've read here. Function is the objective not just books like this and simulation labs. It's about what actually works. Always figure out what is actually best for the situation not just what should be best in theory because theories often have exceptions and new learning to be added. The world around us is always changing. New technologies, new opportunities and new practices are developing all the time and sometimes one of the thousands of other industries might have solved an inefficiency that exists in the way you achieve your objective as well.

For graphics, I expect a lot of new technologies to be coming out even more frequently. Keep looking at technologies in other industries and ask yourself if it could help HMI's be designed better. Make sure it is really making the way you achieve your objective better and not just making the task you perform better by adding more extras or making the interface more appealing or give it more options. Just look at the big picture and ask yourself, "Is the end result better or worse from this?" If it is not better don't do it. But if it could be better please go for it and keep us all safer and more productive.

Thank you for reading *Machine, Meet Human*.
I hope you found it useful and will want to pass it along to others.

Glossary:

AI - Analog Input

An input signals that can have values in a given range, usually as a voltage reading where the number of values possible is relative to the tolerance of the voltage reading.

AO - Analog Output

An output signal that varies the output value with the number of values relative to it's tolerances.

CAD - Computer Aided Drafting

Using computers to aid in designing, drafting and modeling of ideas and replicating the real world.

Curse of Knowledge

The concept that people who attain a certain level of understanding about a given topic often forget all the things they learned that build up to that understanding. We forget what it is like to not know something and how it affects their perspective on the topic.

153

Example of the curse of knowledge: Explain the meaning of the word "ballad" to someone without using any examples. It is very difficult. The curse of knowledge makes it difficult to view something you know from the perspective of someone who doesn't know it or even before you learned it yourself, even if it was recently.

DCS - Distributed **C**ontrol **S**ystem
System of computers where different computers are responsible for different portions of a process that all work together to monitor and control something.

DI - Digital **I**nput
An input signals that are either On or Off

DO - Digital **O**utput
An output signals that are either On or Off

Gorilla Arm Syndrome -
A condition that occurs when humans are required to keep their arms elevated for more than a few minutes. It results in stiff muscles, lack of fine motor control and general discomfort and fatigue.

GUI - Graphical User Interface

A visual screen responsible for providing information to a user by graphically representing it. i.e. a computer monitor.

HCI - Human Computer Interface

An HMI specific to the interaction between a human and a computer.

HMI - Human Machine Interface

Any application where humans interact with Machines. This includes computers work stations, airplane cockpits even using a blender involves human machine interfacing.

IO - Input / Output

signals and variables transmitted between the computer and any hardware outside of the computer.

MMI - Man Machine Interface

Same as an HCI

P&ID - **P**iping and **I**nstrumentation **D**iagram
Technical schematic drawings showing how everything is connected and what sizing everything is, even though the drawing is generally not to scale.

PLC - **P**rogrammable **L**ogic **C**ontrollers
Single processor basic programmable computers usually installed in the field that carry out much of the local control and receive commands from the SCADA about what their objective is and they have their own logic to achieve that objective. Often can be stand along systems.

RTU - **R**emote **T**erminal **U**nit
An HMI that is located remotely, generally out in the field near the equipment, but separated from the rest of the control system computers and main SCADA control room.

SCADA - **S**upervisory **C**ontrol **A**nd **D**ata **A**cquisition
System of computers and associated hardware that gathers information about the field equipment, makes decisions and controls the plant according to it's programming.